# Sustainable Human Resource Management: Policies and Practices

# RIVER PUBLISHERS SERIES IN MANAGEMENT SCIENCES AND ENGINEERING

*Series Editors*:

**CAROLINA MACHADO**
*University of Minho*
*Portugal*

**J. PAULO DAVIM**
*University of Aveiro*
*Portugal*

Indexing: All books published in this series are submitted to the Web of Science Book Citation Index (BkCI), to SCOPUS, to CrossRef and to Google Scholar for evaluation and indexing.

The "River Publishers Series in Management Sciences and Engineering" looks to publish high quality books on management sciences and engineering. Providing discussion and the exchange of information on principles, strategies, models, techniques, methodologies and applications of management sciences and engineering in the field of industry, commerce and services, it aims to communicate the latest developments and thinking on the management subject world-wide. It seeks to link management sciences and engineering disciplines to promote sustainable development, highlighting cultural and geographic diversity in studies of human resource management and engineering and uses that have a special impact on organizational communications, change processes and work practices, reflecting the diversity of societal and infrastructural conditions.

The main aim of this book series is to provide channel of communication to disseminate knowledge between academics/researchers and managers. This series can serve as a useful reference for academics, researchers, managers, engineers, and other professionals in related matters with management sciences and engineering.

Books published in the series include research monographs, edited volumes, handbooks and text books. The books provide professionals, researchers, educators, and advanced students in the field with an invaluable insight into the latest research and developments.

Topics covered in the series include, but are by no means restricted to the following:

- Human Resources Management
- Culture and Organisational Behaviour
- Higher Education for Sustainability
- SME Management
- Strategic Management
- Entrepreneurship and Business Strategy
- Interdisciplinary Management
- Management and Engineering Education
- Knowledge Management
- Operations Strategy and Planning
- Sustainable Management and Engineering
- Production and Industrial Engineering
- Materials and Manufacturing Processes
- Manufacturing Engineering
- Interdisciplinary Engineering

For a list of other books in this series, visit www.riverpublishers.com

# Sustainable Human Resource Management: Policies and Practices

**Editor**

**Carolina Machado**

University of Minho
Portugal

LONDON AND NEW YORK

**Published 2019 by River Publishers**
River Publishers
Alsbjergvej 10, 9260 Gistrup, Denmark
www.riverpublishers.com

**Distributed exclusively by Routledge**
4 Park Square, Milton Park, Abingdon, Oxon OX14 4RN
605 Third Avenue, New York, NY 10158

First published in paperback 2024

*Sustainable Human Resource Management: Policies and Practices* / by Carolina Machado.

*Routledge is an imprint of the Taylor & Francis Group, an informa business*

Publisher's Note
The publisher has gone to great lengths to ensure the quality of this reprint but points out that some imperfections in the original copies may be apparent.

While every effort is made to provide dependable information, the publisher, authors, and editors cannot be held responsible for any errors or omissions.

ISBN: 978-87-7022-120-7 (hbk)
ISBN: 978-87-7004-350-2 (pbk)
ISBN: 978-1-003-33966-3 (ebk)

DOI: 10.1201/9781003339663

# Contents

# Preface

This book looks to cover the issues related to sustainable human resource management policies and practices in a context where organizations are facing, day after day, high challenges related to the continuous changes in the markets as well as in the environment as a whole. Nowadays, organizations need to be increasingly strategic in their actions, developing efficient and effective policies and practices in order to more effectively overcome those challenges. More than just react, these policies and practices need to be increasingly proactive. At this level, human resource management is not an exception. On the contrary, considered as one of the most critical resources of an organization, it is crucial to develop effective human resource management policies and practices as a way of obtaining innovative and creative human resources in the organization. Issues such as corporate social responsibility, organizational citizenship, innovation, creativity, organizational diversity, internationalization, ethics, governance, and sustainability are increasingly present in organizations that look to be strategic and competitive in the actual strongly competitive markets. Conscious of these challenges and their impact on human resource management policies and practices that competitive and responsible organizations need to develop, this book assumes a critical relevance to all those that as researchers in this study area or as managers in the different kind of organizations need to manage and develop their human talents in a sustainable way, contributing to have effective human resources not only today, but also more and more in the future.

With this book, we look to increase the discussion and debate about the role of human resource management in developing sustainable work practices as well as sustainable human resource management systems. The role of human resource management, namely in what concerns its policies and practices, is of critical relevance in supporting the organization's sustainable business.

The organization's management and engineering areas play here an important role as they need to act ethically and in a sustainable way, contributing to the organizational development; at the same time, they need to improve and contribute to the workforce well-being, as well as the society quality of life, at present and in the future.

Taking into account these concerns, this book looks to cover the issues related to sustainable human resource management policies and practices in a context where organizations are facing, day after day, high challenges in what concerns the items related to sustainability. It looks to provide a support to academics and researchers, as well as those operating in the management field need to deal with policies and strategies related to human resource management and sustainability.

Following its main aims, this book looks to cover the field of *sustainable human resource management policies and practices* in five chapters. Chapter 1 *"Human Resource Systems for sustainable Companies"*; Chapter 2 *"Managing Green Recruitment for Attracting Pro-environmental Job Seekers: Toward a Conceptual Model of 'Handicap' Principle"*; Chapter 3 *"Sustainable HRM: How SMEs Deal With It?"*; Chapter 4 *"The (Un)sustainable Process of Devolution of HRM Responsibilities to Line Managers"*; and Chapter 5 *"Transversal Competences: A True and Effective Support to Achieve Greater Organizational Sustainability"*.

Understood as a critical tool for academics, researchers, human resource managers, managers, engineers, and other professionals in related matters with human resource management and sustainability, the interest in this subject is evident for many types of organizations, namely, important institutes and universities worldwide.

The editor acknowledges River Publishers for this opportunity and for their professional support. Finally, she thanks all chapter authors for their interest and availability to work on this project.

**Carolina Machado**
Braga, Portugal

# List of Contributors

**Ana Luisa Silva,** *School of Economics and Management, University of Minho, Portugal; E-mail: analuisalobarinhas@gmail.com*

**André Filipe Barreira,** *School of Economics and Management, University of Minho, Portugal; E-mail: barreirandre@gmail.com*

**Carolina Feliciana Machado,** *School of Economics and Management, University of Minho, Portugal; E-mail: carolina@eeg.uminho.pt*

**Delfina Gomes,** *School of Economics and Management, University of Minho, Portugal; E-mail: dgomes@eeg.uminho.pt*

**Do Dieu Thu Pham,** *Department of Management, Faculty of Business Administration, Université Laval, Canada; E-mail: do-dieu-thu.pham.1@ulaval.ca*

**João Leite Ribeiro,** *School of Economics and Management, University of Minho, Portugal; Email: joser@eeg.uminho.pt*

**Michael Beer,** *Harvard Business School, USA; E-mail: mbeer@hbs.edu*

**Pascal Paillé,** *Department of Management, Faculty of Business Administration, Université Laval, Canada; E-mail: ppaille72@gmail.com*

# List of Figures

# List of Tables

# List of Abbreviations

| | |
|---|---|
| BD | Becton Dickinson |
| CA | Continental Airlines |
| CEO | Chief Executive Officer |
| CEP | Corporate Environmental Performance |
| CSP | Corporate Social Performance |
| ECWB | Environmental Counterproductive Workplace Behaviour |
| EM | Environmental Management |
| EMP | Environmental Management Performance |
| EMS | Environmental Management System |
| EOP | Environmental Operation Performance |
| ES | Environmental Sustainability |
| GE | General Electric |
| GHRM | Green Human Resource Management |
| GRI | Global Reporting Initiative |
| HCHP | High Commitment, High Performance |
| HP | Hewlett Packard |
| HR | Human Resources |
| HRM | Human Resource Management |
| IM | Impression Management |
| OCBE | Organisational Citizenship Behaviour for Environment |
| P–O | Person–Organisation |
| SHRM | Strategic Human Resource Management |
| SME | Small and medium sized enterprises |
| TBL | Triple-bottom line |
| TPB | Theory of Planned Behaviour |

# 1

# Human Resource Systems for Sustainable Companies

**Michael Beer**

Harvard Business School, USA
E-mail: mbeer@hbs.edu

A human resource system that is not explicitly designed to develop an effective high performing organization with high trust and commitment as well as capacity for change and adaptability will not contribute to a company's sustainability–its capacity to survive and prosper in the long run. In this chapter, I will discuss what kind of HR system helps make a company sustainable, noting right off that the HR system is composed of multiple policies and practices that human resource function is responsible for developing and administering. Their ability to so is, however, as I will discuss, dependent on the company's CEO and senior team's commitment to develop a caring human system as well as high-performing system of organizing, managing and leading. Both of these goals must be achieved simultaneously for sustainability to be achieved.

I will begin with an outline of what makes a company sustainable, then spend some time with an excellent and well-known example–Southwest Airlines. With that foundation established, I will discuss in detail what makes for an HR system that contributes to sustainability.

## 1.1 High-commitment, High-performance Companies are Sustainable Companies

I have made a long study of what I call "high-commitment, high perfor-mance" (HCHP) companies.[1] "High-commitment" means that employees at all levels feel an emotional commitment to the company's goals and values. (This, of course, means that the goals and values are such that all of its stake-holders (employees as well as customer, community, society and of course investors) can feel an emotional commitment to them.) "High performance" means that the company is effective in executing its ever-changing strategies, so it can consistently be a top performer in its industry. (This, of course, means that a whole host of things are being done right, all at the same time.) In fact, these two qualities–high-commitment and high-performance–are not simply two desirable qualities. They are inseparable; they make each other possible.

Although I didn't add "high sustainability" to the name (which is already quite long enough), sustainability is what you get for being a high-commitment, high-performance company. Such companies are able to deliver sustained performance–by staying ahead of competitors through innova-tion, adaptation, and resilience to setbacks–because they have developed the following three organizational pillars (see Figure 1.1):

1. *Performance alignment.* All the elements of the organization–its struc-ture, systems, people, and culture–are carefully designed to achieve its performance goals and strategy. Elements that, for whatever reason, are found to be obstacles to goals and strategy are changed accordingly rather than left alone or worked around.
2. *Psychological alignment.* Employees at all levels–particularly key unit leaders–are emotionally attached to the organization's purpose, mission, and values. People truly care about these things.
3. *Capacity for learning and change.* Employees at all levels, starting with its leaders are able to have honest collective (involving all key people) conversations about internal realities that require change necessary for the company sustain its commitment and performance over time. It is these conversations that enable the organization to learn from its

---

[1]For a full discussion of high-commitment, high-performance companies, see my book *High Commitment, High Performance: How to Build a Resilient Organization for Sustained Advantage* (San Francisco: Jossey-Bass, 2009, and Higher Ambition: How Great Leaders Create Economic and Social Value (Boston, Harvard Business Publishing, 2011).

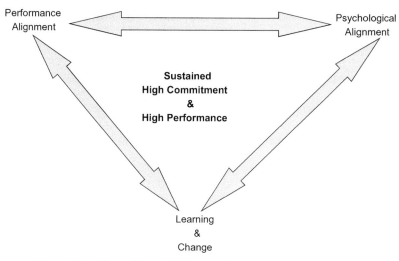

**Figure 1.1** Three pillars of high commitment and high performance.

successes and failure to effectively execute its strategies and liver to its declared human values.

It is not enough for a company to have one or two of these pillars in place at a given time. As a *system*, all three must be in place and strong at the same time. That is what enables the virtuous cycle depicted in Figure 1.2 and endows an organization with resilience to weather the inevitable ups and downs of business–that is, makes it sustainable.

A key part of any HCHP company is its distinctive HR system. An HCHP company must attract, hire, train, socialize, compensate, and promote (or terminate) its people in ways that create and maintain the three pillars. This sometimes calls for HR practices that are not the norm or that seem to many counterintuitive in a crisis.

Can such a company really exist? Yes, it certainly can. I am not describing an ideal. I am describing companies that I and other scholars have closely observed. Table 1.1 provides an illustrative list of high-commitment, high-performance companies in industries as diverse as retail, steel, education, sports, and high tech. All have outperformed their peers for many years–in some cases, for decades.[2]

---

[2] Analysis by Beer of these companies confirmed that their annual compounded growth in revenues, profits and market capitalization is above average. Cases, books and articles were used to evaluate each company's culture.

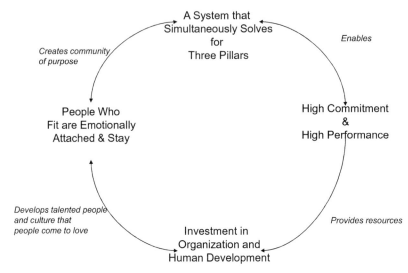

**Figure 1.2**    The virtuous cycle.

**Table 1.1**    High commitment and high performance companies

| | |
|---|---|
| Airlines | Southwest Airlines |
| Software | SAS Institute |
| Technology | Hewlett Packard (1937–1999) |
| Telecommunication | Cisco |
| Consulting | McKinsey |
| Medical Products Distribution | Henry Schein |
| Sports | New England Patriots |
| Steel | Nucor |
| Pharmaceuticals | Johnson |
| Hotels | Marriott |
| Medical Products | Becton Dickinson |
| Electrical Equipment | Lincoln Electric |
| Grocery | Costco |
| Education | Harvard Business School |
| Automobiles | Toyota |

To gain a detailed understanding of what one of these companies look and feel like, let us take a careful look at Southwest Airlines. It makes for an excellent illustrative case, not only because it has been so successful for so long, but also because it has been so well researched, and its policies and practices have been so well documented. It is clearly a sustainable company and its HR system can show us what goes into a human resource system for a sustainable company.

### 1.1.1 Southwest Airlines: An Illustrative Case

Southwest Airlines, founded in 1971 with only three airplanes, has been the most successful airline company in the United States for 48 years [1–3]. As of 2018, it employed 58,000 people. Its annualized returns to stockholders has been 17.5% compared to 11% for the broader market [4, 5]. Southwest has never had to declare bankruptcy in an industry in which every airline has had to do it–that is sustainability. Since 1971, it has outperformed all other airlines on a number of critical operating dimensions, and is the only airlines to have made a profit every year for the past 48 years. Southwest's operating costs are 20% below the industry average, despite the fact that it is 80% unionized. Most observers regard the airline industry, deregulated since the late 1970s, as a very tough one–even a terrible one in which to compete. Yet Southwest's returns exceed those of many companies in industries considered much easier to succeed in.

Southwest's pulls all this off with a well-defined strategy and a set of very counter-conventional management policies and practices. These, in turn–as we will see–rely on an HR system designed to support them.

Southwest's original strategy, to serve non-business customers with low fares that would compete with bus fares on that route, was thought at the time to be impossible and frankly crazy. Southwest has since appealed to business travelers and found that it can attract them as well with its low cost and friendly service. It has worked, though: Southwest has been able to maintain low fares. They did it–and continue to do it–by creating a community of people who relentlessly focus on controlling cost; that is, a high-commitment company. Everyone at Southwest takes this seriously. Everyone cares. Everyone is willing to help accomplish this, even if it is not to their individual advantage. Here is how Colleen Barrett, former Executive Vice President of Customers, explained it:

> We're one big family, and the family members expect a lot of each other. Part of that means watching our costs carefully. We can't compete unless our costs are as low as, or lower than our competitors, which mean everyone has to take part. For example, when the price of jet fuel skyrocketed during the Persian Gulf crisis, Herb [Kelleher, the CEO of Southwest during its first three decades] sent a letter to every pilot asking if she or he would contribute ideas on how to save fuel. The pilots developed a new procedure for takeoffs and landings that was just as safe and saved a significant percentage of the fuel used in those maneuvers.

Gary Kelly, Southwest's current CEO has continued to enable this high involvement approach to managing.

Southwest Airlines also developed an operating system that supported its low-price, low-cost, on-time departure strategy. This model broke every rule in the airline industry at the time. It includes flying only one type of aircraft (the Boeing 737), fast turnaround of airplanes, timely departure, high utilization of airplanes, flexible work roles, and a friendly and fun environment for customers that keeps them coming back. Historically, the company flew out of secondary, uncongested airports, though by 2018 it also flies out of major airports and flies long haul routes a departure from its original strategy and offers point-to-point service only; passengers cannot connect to other cities through the traditional hub-and-spoke system. And in contrast to the practice of other airlines at the time, Southwest Airlines does not provide assigned seats, serves only peanuts and drinks, and does not check luggage through to another airline when a passenger wants to make a connection.

Despite not doing all these things that customers supposedly had to have done for them and despite conventional wisdom that low cost means less personal service; Southwest customers love flying Southwest. The company earned the Triple Crown award–best on-time performance, fewest complaints, and fewest bags lost–for 5 years in a row, something that no other airline has achieved. It ranked number one for the fewest customer complaints for 13 consecutive years. And it continues to excel in all these outcomes.

This was accomplished not with technology but with a human *touch*; passengers like flying Southwest for the fun and friendliness. As one flight attendant stated, "We are all dedicated to the delivery of positively outrageous service to customers–with a sense of pride, warmth and friendliness" [6]. As founder and first CEO Herb Kelleher liked to point out, "It's easy to offer great service at high cost. It's easy to offer lousy service at low cost. What's tough is offering great service at low cost, and that's what our goal is" [7].

The secret to Southwest's one-two punch of low-cost and friendly service is that its approach to managing people–its HR system–is aligned with its own goals and strategy. One employee who had worked at several other large companies said, "I was pretty dubious at first, having been at places where everyone but the top two or three people were considered commodities, but I have come to appreciate a place where kindness and the human spirit are nurtured" [8]. An industry expert concurred: "At other places, managers say that people are the most important resource, but no one acts on it. At Southwest, they have never lost sight of that fact" [9] Southwest turns up consistently in *Fortune*'s annual list of the "100 Best Companies to Work for in America".

One aspect of company sustainability is the ability to face tough competitors and win. By this token, Southwest is quite sustainable as the two examples in its history exemplify. United Airlines decided that Southwest was threatening its California market, which accounted 45% of its revenues, and tried to compete on the San Francisco–Los Angeles route, only to fail miserably. Continental tried to compete by offering lower-cost "Continental Lite" flights–peanuts and drinks only–but failed to make it pay and gave it up. That is because these airlines could not replicate Southwest's human touch culture, which it takes years to develop, the reason it is the source of sustained advantage.

### 1.1.2 Three Pillars of Sustained HCHP at Southwest

Let us look in more detail at how Southwest maintains the three pillars of a high-commitment, high-performance organization. Then, we will look at the role of the HR system in keeping those three pillars standing.

**Performance Alignment:** High performance requires that *all aspects of the organization's design are aligned or fit together and also are aligned with or fit the strategic task of the organization.*[3] This double requirement in turn requires that management is consistent in the design levers it uses throughout the organization. Why is this so important?

- Strategy Must be Distinctive, Focused, and Values-based: This is a key to competitive advantage. Southwest was the first airline to realize that there was an entirely unserved market for the allegedly impossible combination of cheap air transportation *and* friendly service, although by 2018 several other airlines have come closer to replicating Southwest, notable Jet Blue founded by a former executive at Southwest. Southwest has focused relentlessly on this strategy, which included not serving luxury travelers who wanted other things that could not be provided at

---

[3]The idea that various aspects of the organizational system must be configured to align or fit each other, and the environment/strategy has a long history in the organizational studies literature. It has been shown in many diverse studies as an essential ingredient for effective performance and survival. See for example Miller, D. (1986) "Configurations of Strategy and Structure: Towards a Synthesis," *Strategic Management Journal*. 7, pp. 233–249; Miller, D. (1987) "The Genesis of Configuration," *Academy of Management Review*. 12, pp. 686–701; Miller, D. (1990) Organizational Configurations: Cohesion, Change and Prediction, *Human Relations*. 43, pp. 771–89.

low cost (or at least, not at low cost *with friendly service*). As Gary Kelly, current CEO of Southwest, told me:

[The founders] were remarkably disciplined in focus with the strategy over decades and resisted the temptation to chase other distractions . . . They knew what they were good at and they just worked harder and harder and harder over the years to perfect that.[4]

Such a focused strategy made it possible for management to create a well-aligned system and to hire people who, as we saw in the quotes above, were aligned with their strategy of low cost and friendly service. The organizational design, the operating system, and the human resource policies were all specifically designed to allow an airline to compete with car, bus, and train travel. It is equally important, though less often recognized, that HCHP firms' strategic choices emerge not only from their heads but from their hearts. That is, the strategy reflects a rational assessment of market opportunities, but it also must reflect senior management's desire to make the world a better place. It is this value-based aspect of the strategy that enables the company to avoid distractions, as Kelly put it. It means that one way of making money *isn't* necessarily as good as any other. Gary Kelly sees human values as the underpinning of Southwest's strategy.

A huge component of the strategy wasn't necessarily a thought. It was more visceral. It was just an appreciation for people. It was, I think, a very deep understanding that we're here to serve customers and that, in order to serve customers well, you have to have great people and you have to take great care of them.

• The Organization Must Fit the Strategy: Without a clear strategic choice, it is impossible to fit the organization and its human resource system to the strategy. By organization, I mean the roles, responsibilities, and relationships that make specialization and coordination possible. Structure and systems can define these formally, but in HCHP companies, the culture and norms play an even more important part in shaping roles, responsibilities, and relationships. Organized human efforts presumably have to go somewhere. If they are not all going towards the same organizational strategy, then some of them are going against it (if only by wasting some of the total effort) and some are going against

---

[4]Interview with Gary Kelly, May 14, 2008.

each other. Thus, without alignment, both strategy and performance will be undermined. Even if a poorly aligned company manages to have decent performance, it will be vulnerable to a company whose aligned performance is much better. Percy Barnevick, former CEO of ABB, has said that strategy–or thinking–is 10% of the battle, while implementation–which includes alignment–is 90%."[5]

Southwest's strategy, for example, requires 15-minute turnarounds for its airplanes and a nonstop focus on keeping cost low. The operating system was designed (that is, *purposely*) to be aligned with this strategy. Southwest uses only one type of plane, which makes for lower maintenance costs. It does not offer assigned seating, which drastically simplifies–and lowers the costs of–its booking and boarding operations. Southwest created the position of airport station manager, which no other airline had. The station manager is responsible for molding the maintenance staff, pilots, ramp workers, and customer service employees into a tight-knit, cross-functional team, which, in turn, is what makes it possible to turn a plane around in only 15 minutes. He or she is empowered to lead that team in a way that deals best with the local airport conditions, rather than being driven by rules and processes handed down from corporate staff who may or may not know much about airports but, in any case, do not know the day-to-day situation at any particular airport. Southwest's team-driven approach not only is more in tune with local conditions, but also encourages–in fact, it demands–initiative and teamwork. These, in turn, create buy-in and dedication; employees almost always do better work when they see their job more as getting the job done than as following the rules. And of course, well-tuned teams tend to be more efficient–that is, more cost-effective–that is, aligned with the company's overall strategy.

On the people side, Southwest's low-cost, fun-and-friendly service strategy demands (a) managers who are not elitist or particularly concerned about status and (b) employees who are relationship-oriented and relatively uninhibited. Even pilots have to have these qualities, no matter how good they are at flying the planes. Pilots who are not relationship-orientated would not collaborate with flight attendants in cleaning an airplane or collaborate with the ground crew to meet the 15-minute turnaround objective. Nor would they be able to relate to passengers in the friendly manner expected of *all* Southwest employees; customer loyalty depends on customers having a

---

[5] Video: *Barnevik, Percy and ABB*. (1994) Produced by Mafred F.R. Kets de Vries. Boston: Harvard Business School, Media Services Library.

consistent experience with the company. In short, a super-skilled pilot who just wanted to fly the plane and leave fun-and-friendly part to the flight attendants would be dangerously unaligned with Southwest's strategy and its value proposition to its customers.

High-performance organizations like Southwest are clear about the attitudes, capabilities, and behaviors needed to implement their strategies. They therefore require and create an HR system that can recruit and select people who fit these criteria. A corollary is that they are clear about qualities they do not want; they therefore require and create an HR system that can spot and avoid people who have such qualities. A highly proficient pilot who applied for a job at Southwest was turned down because he was rude to a secretary. The HR system was set up to take an observation like that seriously. That pilot probably would have been hired by a non-HCHP company, on the assumption that it makes sense to hire candidate with the best functional skills as opposed to one with the right attitudes for and fit with the company culture. Such hiring mistakes are a serious hidden cost for many companies, reducing the value both for customers and for shareholders. In fact, a study of Southwest Airlines shows that coordination based on relationships rather than on structure or incentives is at the heart of that company's high performance [10].

Continental's and United's failure to take away Southwest's customers in the early 1990s can be traced to their inability to align their own cultures and operating systems to the low-cost, friendly-service strategy they were trying to emulate. They were fish who were trying to fly by jumping out of the water and flapping their fins.

- Organizational Design Levers Must be Internally Consistent:[6] A workforce can only be high-commitment and high-performance work if people believe that management means what it says–by no means the normal condition. This, in turn, requires internal consistency between the various organization design and human resource levers, the key facets of the organization that shape organizational behavior. Southwest's low cost and rapid turnaround of airplanes would not be possible, as I noted above, without trust and good communication. Consider the following description of how cross-functional communication at SWA has enabled both an aligned and internally consistent organization over the years:

---

[6]The discussion in this section is informed by Barron, J.N. and Kreps, D.M. (1999) *Strategic Human Resource Management.* New York: Wiley, Chapter 3.

> There is a constant communication between customer service and the ramp. When planes have to be switched and bags must be moved, customer service will advise the ramp directly or through operations. If there is an aircraft swap operation keeps everyone informed. It happens smoothly [11].

Consistent policies and practices are also important because inconsistent policies and practices undermine each other's effectiveness. For example, Southwest employees receive extensive training in interpersonal relationships and teamwork at the company's People University. But that training would largely be squandered without an organizational and human resource system that demands (a) accountability for broad performance measures at the station level where it come together, (b) leaders who embody core principles, (c) good union–management relationships, and (d) broad job descriptions. Similarly, a structure that enables station managers to be accountable for broad measures of performance would largely be squandered without station managers who know how to lead a team, a reason why, as I will discuss below, HCHP companies human resource policies must be focused on leadership development.

- Internal Consistency and Competitive Advantage: Internal consistency makes it very difficult for a competitor to copy a firm's formula for success. Copying one practice may be relatively easy; copying a large set of policies and practices that are internally consistent is far less likely. This is in part a function of statistical probabilities [12]. Even if the probability of copying any one policy is .9, the probability of copying many consistent policies quickly becomes far lower. But copying a firm's internally consistent policies is also made difficult by the fact that internal consistency emerges from the CEO's values. A competitor whose CEO does not have the necessary high-commitment, high-performance values cannot create consistent policies and practices. Yet few boards of directors actually take HCHP values into account when hiring a CEO [13]. And it is not just the CEO who needs these values, it is the leaders at every level of the company. Such leadership consistency takes a long time to develop, which, of course, is where the company's whole HR system comes into play and makes emulations so difficult. Southwest achieved this consistency in its leaders' values through many years of careful selection and management development.

United and Continental Airlines failed to compete successfully with Southwest because neither has company-wide leadership with a consistent set of HCHP values. In 1993, for example, Continental's CEO Robert Ferguson "announced his plan to split the company into two operations; one would concentrate on short-haul, low fare flights (named Continental Light or CA Light), and the other would feature first class service at business class prices. Ferguson believed a cost structure lower even than Southwest's would enable Continental to compete successfully" [14]. CA Light cut its fares and cut its meal service, but much of the rest of its operating system–including multiple airplane types, a reservation system, and the hub-and-spoke system– were left in place. More importantly, the company was unable to replicate Southwest's high-commitment culture. Continental flight attendants, when asked to emulate Southwest practice by helping to clean cabins between legs of a flight, complained that their workload had increased, and their breaks were too short. The pilots, too, were also upset until the company provided meals during their busy schedule. Though ground crews tried hard to turn planes around in 20 minutes–that is, *almost* as quickly as Southwest did–they still took 30 minutes.

It takes years to create the Southwest culture and employee attitudes. Perhaps more importantly, Robert Ferguson did not possess the values and style required of a CEO who wants to create a high commitment system. He was described as a taskmaster who was "harsh and uncommunicative" and who drove employees away. He admitted to not suffering fools gladly and said "he would tell you in front of 20 people or 100 people if you were not doing a good job" [15]. This is not the inspiring and caring leadership style of high-commitment leaders like Herb Kelleher.

**Psychological Alignment:** To understand the emotional attachment that defines a firm with psychological alignment, consider this testimony from a Southwest employee: "Working here is truly an unbelievable experience. They treat you with respect, pay you well, and empower you. They use your ideas to solve problems. They encourage you to be yourself. I love working here" [16]. Research has found this to be a typical response to working in a high-commitment system [17]. Employees can be moved in a certain direction–at least for a while–by the "hard aspects" of an organization; that is, its structure, performance management system, and incentives. But employees who are psychologically aligned with the organization's mission and values are *internally motivated*. They form a *community of purpose*. Over and over, employees in such organizations have told me–and many other researchers–that it is the relationships and teamwork, not the rules and regulations (though those are always there) that inspire them to work as hard, as well, and as cooperatively as they do. In that kind of environment, people are willing to sacrifice their immediate self-interests so that their team and their company can achieve high performance–in which they can then feel legitimate pride. As Gary Kelly told me:

> There's a real celebration of success and a celebration of individual achievements and the family-like feel is accomplished through not only what we do here at SWA but also appreciating people for who they are, their personal accomplishment, very simple things like recognizing birthdates and wedding and anniversaries. Many companies do those things. But it comes across, I think, a little disingenuous. But here it really is a celebration.[7]

Firms seeking psychological alignment work hard to create and maintain a distinctive "*psychological contract*"; that is, a high-investment, high-return exchange between the firm (in the person of its managers) and its employees. The psychological contract is, of course, not a formal, legal contract. It may or may not be written down, but it is well understood, and employees know when it has been violated. The psychological contract defines what top management, representing the shareholder, and employees believe they are giving to each other and getting from each other. In a high-commitment culture, there are high expectations and heavy obligations on both sides, but that is what makes such an organization both so effective and so resilient. A key point is that it is based on positive assumptions about what people aspire to and what they are capable of, rather than negative assumptions about their limits and what they might do if they are not held to account [18].

At a minimum, all organizations expect managers and workers to perform their assigned tasks reliably and to achieve performance standards; meanwhile, all employees–from the highest-level managers to the lowest-level workers–expect the organization to treat them fairly, provide fair pay and benefits, and give them at least some say in their own work.

But in high-commitment organizations with high psychological alignment, the demands are much greater in both directions. Management expects that employees will take initiative, work collaboratively, supervise and regulate themselves to a significant extent, work unselfishly, and continually change and adapt. Thus, in an HCHP company, the psychological contract demands of the employees an emotional commitment to the community. Employees, meanwhile, expect much more from the company than fair pay

---

[7]Interview with Gary Kelly by Michael Beer, May 14, 2008.

and fair treatment. They expect management to live by the values that bind the community. They expect open and trusting relationships, achievement, involvement, challenge, responsibility, personal growth, and meaning. It is not that every single human being has these needs, but HPHC organizations select employees with these needs because those are the kinds of employees who are willing to give their all within the right kind of psychological contact.

In effect, the enterprise cedes substantial influence over goals and process to employees in return for dedication and commitment to the community and its purpose. This is what *aligns employees' hearts and minds* with an organization's mission and values. With such a psychological contract, employees have the interests of the company and its customers at heart and are flexible in the work they do (that is, they do not refuse to do something just because it is not in their job description). "We're one big family," Colleen Barrett at SWA explained, "and the family members expect a lot of each other. Part of that means watching our costs carefully. We can't compete unless our costs are as low as, or lower than our competitors, which means everyone has to take part." A ticket agent at Southwest illustrated this commitment to keeping cost low when she pursued a three-dollar stapler she had lent to a colleague at another airline. In most companies, employees would not think that's worth the effort; they might even take the stapler home themselves on the grounds that "the company can afford it." In a psychologically aligned company, people do not feel that it's okay to screw the company now and then.

Table 1.2 summarizes the psychological contract that characterizes a high-commitment, high-performance organization.

**Table 1.2**    The HCHP psychological contract

| Management Expects | Employees Expect |
| --- | --- |
| • Dedication to mission and strategy | • Non-political culture: people do the "right thing" |
| • High performance | • Be on winning team |
| • Behavior consistent with values | • "Leader's behavior consistent with values" |
| • Initiative | • Delegation of authority |
| • Collaboration/team work | • Co-workers who share common values |
| • Self management | • Participation in goal setting |
| • Flexibility | • Autonomy |
| • Unselfish contribution | • Unselfish and egalitarian behavior at top |
| • Openness to feedback and learning | • Opportunity to speak truth to power |
| • Commitment to the firm | • Employee development; above- average pay & benefits; good-faith effort to avoid layoffs |

At its most basic level, Southwest's success can be attributed to the partnerships between senior managers and lower-level managers and between those managers and their thousands of employees. Everyone is involved *as a community* in making the business and their own *joint* work life better. An observation by a United Airlines mechanic in the 1990s illustrates how Southwest differed and undoubtedly reveals one element of United's failure to compete with Southwest:

> You've got to hope that the way we do things will change. [United] is the kind of place where management usually thinks they are way up there and the rest of us are way down here. They want to make sure the shareholders get what they want, but they don't care much about the employees. All that's got to change [if we are going to compete with Southwest] [19].

Employees hope to work for a successful organization, not only for the chance to feel pride and a sense of accomplishment, but also because a successful company is better able to offer personal development, career opportunities, job security, and financial rewards such as stock options. HCHP companies tend to be offer such rewards to employees because the employees' commitment–the product of the psychological contract–produces the financial results that make those rewards available. This is the virtuous cycle represented earlier in Figure 1.2.

The work of developing psychological alignment begins during the recruitment and selection process. Applicants are carefully screened for their fit with the company's values and culture. New hires are clearly told what they can expect and what will be expected of them. Southwest has historically hired only .04 percent of its 202,357 job applicants. Such a selection ratio enables a company to reject employees that do not fit. (Of course, the company would not have so many applicants unless it had first created a reputation as a great place to work.) A socialization process also shapes the psychological alignment. New employees participate in an educational process about the company's values, operating and service principles, and culture. This training is then reinforced through on-the-job experience.

Psychological alignment is particularly vulnerable to inconsistencies in how higher- and lower-level employees are treated. High-commitment organizations typically make a concerted effort to create common policies and practices for upper and lower levels, an approach called "symbolic egalitarianism." Southwest's profit-sharing plan, for example, pays the same percentage of profit to employees at all levels. High-commitment firms typically have open offices with similar cubicles for all levels. They do not bestow status symbols such as cars on their executives.

United Airlines, in contrast, historically created a very different climate, as reported by a flight attendant:

We've always been treated like angry children who don't deserve what they get. Upper management has been adversarial and confrontational with us for over 10 years now; I don't think Mr. Wolf (CEO of United in the early 1990s) liked the flight attendants at all. We are managed differently from other groups. We're disciplined if we're sick more than three days per month or if we arrive late for a flight. We're the only group that has to hop on a scale every month. Pilots certainly aren't held to those standards. When it comes to the boys in the cockpit, things are different. The pilots stay in downtown hotels and we are stuck out at the airport. When we have to deadhead, they fly in first class and we're in the back of the plane. That says it all... The irony, of course, is that the bosses ought to think a lot harder about how we feel if they want to keep their customers happy. We're the people who spend all the time with passengers. To the public, we are United. [20]

**Capacity for Learning and Change:** Robert Bauman, former CEO of Smith Klein Beecham and an experienced change leader, captured the importance of the capacity for learning and change as a key to competitive advantage:

Most important in implementing change in the near term is instilling the capacity for change within the organization in the long term, Bauman said. In my view, the capacity for ongoing change is the ultimate source of competitive advantage [21]. In fact, most firms do not adapt and therefore lose whatever competitive advantage they once had and are unable to sustain their high performance over time, Chris Worley and his colleagues found. Using a benchmark of average industry ROA, they found that only 18% of businesses outperformed the benchmark 80% of the time. Thirteen percent underperformed the industry average 80% of the time. Sixty-eight percent of businesses "thrashed" between under-performance and average performance [22]. In both studies, high performers tended to remain high performers and under-performers remained under-performers. In short, very few companies are able to sustain high performance over time. Many either go bankrupt or are bought out. Worley and his colleagues found that those firms that sustained high performance were more agile–less hierarchical, more collaborative and open to the truth about markets and internal problems, all characteristics of high commitment companies.

Danny Miller, in a groundbreaking study of why successful firms decline, concludes that the very alignment of leadership behavior, policies, and management practices that makes companies successful is also the source of their demise [23]. Successful firms naturally favor and institutionalize the strategy and the management practices that produced that success and suppress divergent views and practices. The powerful become more powerful and defend the company–as they see it–against contrarian views about strategy and adaptation. Also, over time, firms tend to attract, select, promote, and terminate people based on their fit with the firm's historically successful characteristics and practices [24]. This reduces diversity, an essential quality of the work force if firms are to adapt to new circumstances. Eventually, when the right crisis comes along, it proves the Old Guard wrong–but by then, it is often too late [25].

This would seem to pose a big problem for the company's HR system. If a company has been purposely "built for high fit"–that is, staffed with people who agree on certain values and feel committed to the company–can that same company also be built to rebel against itself and change its own stripes? My answer to this question is a qualified yes. It can, so long as it builds honest and fact-based inquiry into its culture. That means a culture in which there are safe ways *at any level* to speak truth to power. Southwest Airline's Herb Kelleher is among minority of CEOs who have been able to create such an environment consistently. Southwest's capacity for change and learning enabled it to adapt from a small Texas company that attracted customers with flight attendants in "hot pants"–to a firm that today has a diverse workforce in varied uniforms. It survived the oil shocks of the 1970s by engaging employees–definitely including non-managerial employees–to solve problems. Southwest scored highest in a study that compared the turnover and return on investment of 10 airlines in the post-9/11 environment, further proof of its capacity to adapt to catastrophic events [26]. Southwest was the only airline that was able to remain profitability over the ten years following the 2008 financial crisis.

What enables HCHP systems to adapt despite the tight alignment of people, organization, culture, and strategy? Even such well-aligned firms are well-supplied with certain kinds of diversity; namely, diversity in functional expertise, diversity that arises from working at different levels, and diversity that arises from interaction with different parts of the firm's environment, such as customers, suppliers, investors, community, and competitors. But few firms have any systematic way to bring these diverse views into an honest dialogue and then make use of the results. In most companies, the hierarchy makes it dangerous–or at least makes it seem dangerous–to tell those higher up that something's wrong and needs fixing. It is usually safer to keep quiet, make do, and let the bosses figure it out for themselves and decide what to do. Often, though, the bosses never do figure it out, or only figure it out too late, or figure it out and propose a solution that those on the front lines know is not going to work.

High-commitment, high-performance firms have a number of interrelated characteristics and practices that create safe circumstances in which senior management and lower levels can (a) engage in such touchy conversations, (b) take action on them, and (c) make sure that action does not just gradually sink into the organizational bog. Here is how they do it:

- Caring About Customers and Performance: In most companies, it is senior management who are responsible for identifying problems and launching initiatives to improve performance, usually with the help of corporate staff groups. Not only are they responsible; they are considered the only ones capable of doing that. In HCHP companies, that assumption is understood to be a willful and fatal blindness. Employees at all levels are expected to care about the customer and about company performance. It is understood that being where they are and doing what they are doing, they in fact know quite a bit about the customers and about company performance that management *doesn't* know. This approach makes a company flexible and much more likely to learn and change; the people dealing head-on with customers and with problems and inefficiencies are often the mostly likely to know how to do better. They are also the most motivated to do so: It is no fun doing something the wrong way when you know you could do it better. It is no fun disappointing customers when you know you could make them happy. This approach has enabled Southwest to overcome some very tough challenges. One customer service supervisor saw it as follows: "The main thing is that everybody cares. We work so many different areas, but it doesn't matter. It's true from the top to the last hired. . . . . . Sometimes my friends ask me, why do you like to work at Southwest? I feel like a dork, but it's because everyone cares" [27]. While the employees' care for their customers is partly their human nature, it is also motivated by their experience that the company cares for them.

Managers and workers who share a genuine focus on their customers and on their company's performance also tend to be uninterested in political or territorial one-upsmanship. That is why a pilot will help the flight attendants get the airplane ready for the next flight.

- <u>Mutual Respect</u>: People are selected for their interpersonal qualities, including both respect for others and willingness to speak up openly and honestly. Once hired, they are encouraged through a variety of mechanisms to exercise these qualities. At Hewlett Packard, for example, management expected–even demanded–that individuals speak their mind. But to make their demand realistic, they created a variety of mechanisms, Including the famous "management by walking around." This was *required* of all managers in order to develop of the trust that would, in turn, make it possible for employees to speak up and tell management unpleasant things it needed to know. Southwest Airlines both encourages and institutionalizes a similar culture [28].
- <u>Egalitarian Culture</u>: As I noted earlier, high-commitment firms tend to be less hierarchical and bureaucratic than their peers [29]. Their egalitarian cultures are free of status symbols such as titles, offices, and private parking spots which makes senior managers much more approachable [30]. As an operations manager at Southwest put it:

    "I would never work at American Airlines. The animosity there is tremendous. Here it's cool. Whether you have a college degree or a GED it doesn't matter. There is no status here, just a good work ethic [31]."

An additional point is that in this kind of environment, diverse ideas are more likely to be heard and become part of the organizational conversation–a key to innovation and adaptability.

- <u>Honest Conversations and Constructive Conflict</u>: Trust makes it possible for managers and workers to have *open and honest conversation* across levels and functional department divides. This, in turn, makes it possible for the organization to solve problems based on facts–*all* the relevant facts, not just the ones that are allowed to filter up to the top in many organizations. The criterion by which possible solutions are judged is what is best for the firm and its customers rather than what is best for one's own career or department.

This attitude has to begin at the top or it would not survive at lower levels. At Southwest, Herb Kelleher exemplified this kind of openness. A pilot observed:

> I can call Herb today. You just call and say there's a problem. He'll say, "Think about it and tell me the solution that you think will work." He has an open-door policy. I can call almost 24 hours a day. If it's an emergency, he will be back to me in 15 minutes [32].

Research shows that in uncertain and turbulent environments, where learning and change is essential for survival, high-performing companies are differentiated from low-performing companies by a culture that encourages employees to confront the inevitable conflicts that arise among functional departments [33, 34]. In HCHP companies, these conflicts occur just as they do anywhere else and they are not swept under the carpet, but neither are they allowed to develop into win–lose arguments. Rather, they are resolved through problem solving, which of course solves problems but also helps maintain the trusting relationships. Here is how some Southwest employees see it [35]:

> What is unique about Southwest is that we are really proactive about conflict. We work very hard at destroying any turf battles once they crop up–and they do. Normally they are not malicious or ill intentioned. Sometimes it's a personality conflict. Sometimes it's bureaucracy.
>
> Station Manager, Southwest Airlines

> Because we are moving at a fast pace, miscommunication and mis-understanding happen sometimes. We take great pride in squaring it away as quickly as possible. Pilots and flight attendants–sometimes an interaction didn't go right between them. They are upset, then we get them together and work it out, it is a teamwork approach. If you have a problem, the best thing is to deal with it yourself. If you can't, then we take the next step–we call a meeting of all the parties.
>
> Pilot, Southwest Airlines

Trust and open dialogue are essential to a HPHC company, but they are also fragile. They can easily fall prey to a conflict over fundamental strategic issues that threaten to change or overthrow the company's historic business model. As I noted earlier, high alignment can ossify and keep the organization from reinventing itself when it needs to. The way to keep this from happening is to create a learning culture; that is, one that institutionalizes a means for honest collective and public conversations, so management can learn from those below them about gaps in performance and psychological alignment before these become extremely difficult or impossible to fix [36].

Southwest, for example, has a culture committee, composed of employees from all levels and parts of the company, who are responsible for assessing all company initiatives for fit with the culture. It also has an open-door policy that makes it practical for employees to skip levels if they need to bring up a problem or complaint that is too hot to take to their own boss. For many years, Lincoln Electric had an Advisory Committee to the CEO which gave employees some influence on the work practices and the incentive structure on which the company's high productivity was founded. Becton Dickinson has institutionalized a *learning and governance process* that enables regular honest and collective conversations about the organization's effectiveness and whether it is living its values and employed it successfully in 2011 to lean how they could become a more innovative and faster growth company [36]. Feedback from the organization identified, among several factors, human resource policies and practices that needed to change if sustainability was to be preserved.

Though the firms mentioned in this section are clearly more adaptive than the average firm, there is also evidence that, even for them, learning and adaptability is the least developed of the three pillars of high commitment and high performance (the other two being performance alignment and psychological alignment). Hewlett Packard, an iconic most admired high-commitment, high-performance company for five decades, went into a serious decline after 1999. Learning requires multiple organizational levers of the organization–the board, the CEO, the top team, and lower-level leaders–to engage in fact-based inquiry, reflection, diagnosis, and then action planning. These processes threaten some people's self-esteem and even their careers unless these processes are designed to ensure anonymity and protect those provide feedback [37].

Southwest began to develop a system for learning in 2005 in response to large and threatening changes in the industry. Gary Kelly described the changes in strategic management he initiated within the senior team and with the larger extended group of senior managers:

> We changed the way we worked dramatically. We have much more frequent interaction. We have much more formalized planning sessions where we all . . . You know, it's not just walking with a whiteboard. But it's basically making sure that we give all of our officers a seat at the table to understand what the challenges are, where we want to be, what it is we're going to have to do to close the gap, so they can lead their area and be a great team member as well. Plus, I wanted to hear what they had to say. So, it's been a very empowering effort for us.[8]

Kelly also noted that this process "required some patience and it requires some nurturing, it requires some vulnerabilities. . . . . . You got to be honest and say, 'You know what? You're right, I don't understand.' Or, 'You're right, the way we did this wasn't as good as we could do it.' Because most people are going to be defensive and reject overtures to maybe do things a little different."

Southwest's capacity for learning and change has enabled it to adapt to numerous changes in the industry. In 2010, for example, with oil at a very high $140 a barrel, Southwest was still profitable, though not as profitable as it had been before.

### 1.1.3 Leader Values and Philosophy

High-commitment, high-performance systems arise from the founders' or the current CEO's philosophy and values; it is unlikely that they can be copied by another company. If it were that easy, Southwest would long since have been awash in successful imitators. All the same, high-commitment, high-performance leaders tend to have certain values in common. Consider the philosophy of Southwest's original and longtime CEO, Herb Kelleher:

> We've never had layoffs, . . .we could have made more money if we furloughed people. But we don't do that. And we honor them constantly. Our people know that when they are sick, we will take care of them. If there are occasions or grief or joy, we will be there for them. They know we value them as people, not just as cogs in a machine [38].

---

[8]From an interview with Gary Kelly conducted by Michael Beer in May 2008.

The CEO of another HCHP company, PSS World Medical has similar values:

> Business people don't like to talk about values. But without these, all business is about is making money.... To me, achieving business goals is great. But no business goal is worth sacrificing your values. If you have to treat people poorly or cut corners in your dealing with customers, forget it...You can build an organization based on mutual loyalty, even in today's economy. But you can't do it if you treat people as disposable [39].

Without such values, a CEO will not be able to make internally consistent business and human resource decisions and will not be able to keep on the high-commitment, high-performance path when things get rough; for example, when capital markets are pushing for easy wins like cutting cost or acquisitions and mergers.

A large-scale Stanford study of high-tech start-ups clearly shows that the founder's basic values and his or her notion of what makes a firm a success shapes the Human Resource system that firm ultimately adopts [40]. Start-up CEOs who thought people were central to success developed high commitment organizations and these were more successful in comparison to others on a number of outcomes. But even leaders who start with an HCHP philosophy have to be open to feedback about the consistency of their decisions and actions with that philosophy. Kelleher, as noted earlier, mandated an open-door policy at Southwest and he really did heed what employees told him.

Kelleher was also the lowest-paid CEO among those of high-performing companies and showed no wish to set himself off from the *community of employees* either in his behavior or in the physical trappings of his position. He was always happy to stop and talk to employees, listen to their problems, and invite them to help solve the problems. He once turned up in "drag" at a maintenance hangar at two in the morning. These were genuine expressions of his values, which included treating employees with care and respect (just as he wanted them to treat the passengers) and helping employees have a good time doing their jobs (as he wanted them to make flying fun for the passengers).

## 1.2　The HCHP Human Resource Management System

A truly sustainable company in our time needs to be what I have described as a high-commitment, high-performance company, with not one or two but all three pillars strong. And as I have been hinting all along, creating and maintaining such a company very much depends on its human resources *system*, by which I broadly mean all the policies and practices, formal and informal, that control or influence how employees interact with each other and with the company's customers and other external parties. Note that some of these policies and practices are in the domain of the organization's Human Resources Department, but others are in the domain of the CEO and his or her executive team.

The question, then, is: What type of human resource policies and practices allow Southwest to have the kinds of people it does and allow those people to do what they do? And more importantly for you, what type of human resource policies and practices would allow your company to have the kinds of people it needs to have and allow those people to do what they need to do in order to be a high-commitment, high-performance *sustainable* company? Although HCHP firms can be found in quite a variety of industries, they are remarkably similar in their HR policies and practices.

A fundamental reason for this is that management makes positive and optimistic assumptions about people. They believe that most people are capable of learning, are motivated intrinsically, and want to contribute and make a difference. With that view, it does not make sense to try to control employees' behavior mainly through monitoring and incentives; that is a bit like taking schoolchildren out to the playground and then ordering them to play, with rewards for who plays the most. Barrett Joyner, a senior executive at software developer SAS Institute–an HCHP company–offers a good sample of HCHP thinking: "The emphasis [at SAS] is on coaching and mentoring rather than monitoring and controlling. Trust and respect–it's amazing how far you can go with it" [41].

No single HR policy or practice can by itself change a company. All of the policies and practices discussed below must work together to contribute to performance alignment, psychological alignment, and capacity for learning and change. (Which is to say, the HR policies and practices, like all the company's other policies and practices, must be aligned with the company's goals and values.)

## 1.2.1 Long-term Employment

Most HCHP companies believe in hiring for the long term–even for life. They see the development of people who fit their culture as an investment they do not plan to liquidate unless absolutely forced to by circumstances. Furthermore, they take steps to avoid such circumstances.

In 2019 most senior executives believe that full employment (job security in time of economic decline) is a do-good policy whose time is up because, with such rapid change, it no longer makes good business sense to provide a guarantee that there will not be layoffs. While this may be true for some companies in very volatile industries, HCHP companies like SAS Institute and Southwest Airlines disagree; they avoid layoffs very much for business reasons. That is not to say that no one ever gets fired; they let low performers go, but they do it humanely and only after due process (performance discussions). For that matter, HCHP companies are more aggressive than others about getting rid of low performers and are cautious about hiring precisely because they know they will try not lay people off in bad times. Although employment security may seem like a policy that would naturally lead to low performance– wouldn't people start to slack off once they knew it wouldn't cost them their job?–the evidence says otherwise. Companies as varied in their industries and business models as Lincoln Electric (welding and electrical equipment), Nucor Steel (mini-mills), Hewlett Packard (technology), New United Motors (automobiles), and Southwest (airlines) have outperformed their competitors for long periods during which they offered employment security and, specifically, did not conduct layoffs. This was not just because times were good; they stuck as much as possible to that policy even when layoffs seemed the obvious step to take. Southwest's Herb Kelleher believed that such a policy means giving up some profits but in return gaining partnership with employees (social capital). "Our most important tools for building employee partnership are job security and a stimulating environment. . . Certainly there are times when we could have made substantially more profits in the short term if we had furloughed people, but we didn't" [43].

Evidence shows that striving to avoid layoffs makes economic sense. A study by Wayne Cascio "found no significant, consistent evidence that employment downsizing led to improved financial performance (stock price and return on assets)" [44]. He concluded that stabilizing employment is sustainable because it improves *performance* as well as commitment. Though there is no way to know whether a long-term employment policy was the cause of high performance or enabled by the company's already high performance. Casio's study does, however, show that profits and a long-term employment policy go together. Interestingly, even in the successful performance turnarounds studied by Jim Collins, "companies rarely used headcount lopping as a tactic and almost never used it as a primary strategy" [45].

How does long a term-employment employment policy lead to high performance? First, losing people due to layoffs or turnover means losing commitment and capabilities that have been developed over many years. Second, frequent or large "in and out employee flows" not only disrupt specific relationships but also make it seem less worthwhile to invest in work relationships in general. That, in turn, undermines employee productivity. And high turnover hurts customer relationships and loyalty. Meanwhile, layoffs increase recruitment and hiring costs, and these often outweigh the payroll savings obtained from layoffs. In a study of manufacturing plant layoff practices at Corning Glass Works in the 1970s, Jim Thurber demonstrated this very effect: plant gross margin would have been approximately the same over the course of the fixed time frame studied without layoffs [46]. Long-term employment develops a long-term perspective among employees and managers, which, in turn, enables employees to take personal risks to improve efficiency without fearing for their security and career path. They can be more flexible and innovative. Managers become more concerned with leaving a successful legacy than with making the next quarter's numbers. A study of family-owned or family-dominated companies with this employment ethos outperformed publicly traded companies on the average [47]. But publicly traded HCHP companies can also reap the benefits of the long-term perspective if they are careful in their policies and strategies regarding risky financial debt and fast growth. By minimizing risks associated with these strategies, they avoid placing the firm in a position in which layoffs become the only way to survive. As Herb Kelleher observed, this may mean forgoing some short-term profits [48]. CEOs who took this approach would certainly not have made the reckless gambles that led to the Wall Street meltdown–and massive financial industry layoffs–of 2008.

Finally, long-term employment forces management to be careful about whom and how many people the company hires. (As a child at summer camp, I learned that when you pack for a canoe trip, remember that everything you take, you will have to carry *on your back* over the roughest portage you might have to make.) Kelleher put it this way:

> It turns out, providing job security imposes additional discipline, because if your goal is to avoid layoffs, then you hire very sparingly. So, our commitment to job security has actually helped us keep our labor force smaller and more productive than our competitors [49].

That's all very well, but does it really make sense in today's turbulent markets to have a long-term employment policy, even an employment security policy? That depends. If senior management does not have the values and philosophy that must underlie such a policy, then no, it won't make sense. If senior management does aspire to adopt a long-term employment policy, the next step is a careful study of the business and of the *fully burdened costs* of the current hire and fire policy. For example, how volatile has the revenue stream been historically? How much is it a function of market forces over which the firm has little control and to what extent can volatility be reduced with more conservative business policies? (Hewlett Packard, during its first few decades, made a deliberate choice to forgo defense contracts, not out of any pacifistic leanings but because they did not want to expose themselves to the defense industry's rather dramatic cycles given their intention to build a high trust and commitment culture. How would recruiting, selection, customer dissatisfaction, and training costs–to name only a few–be reduced with a long-term employment policy? What are the anticipated benefits of long-term employment in terms of added flexibility, commitment, and other benefits discussed earlier?

There is yet another important ethical reason for an employment security policy and that is the stress it causes and its consequent negative impact on physical and mental health [50].

Clearly, it would help a lot if the capital markets adopted a longer time horizon. That would require changes in national policy, for example, policies regarding short- and long-term capital gains. In any case, it is not something an individual company can control. But CEOs of want to build a sustainable company have learned that instead of responding to Wall Street analysts' inquiries about quarterly earnings it is best to start the conversation with investors by defining their long-term strategies, the initiatives they are pursuing in support of their strategies and refusing to provide quarterly guidance – committing to quarterly earnings numbers. Committing to short-term performance has been found to be associated with myopic CEOs and lower long-term performance, so why do it [51].

## 1.2.2 Selective Recruitment and Hiring

HCHP companies are distinct in their insistence on hiring people who fit their culture's values and principles. They look for attitudes, values, and potential to grow and develop, rather than strictly at specific skills and experience, though of course those matter and may even by decisive for essential specialists in niche roles. Research by Jenny Chatman has found that values-based hiring helps create a company in which new employees adjust more quickly and stay longer–enhancing productivity on both counts [52].

Hewlett Packard, in its first five decades, recruited the best engineers from Stanford, but that wasn't a sufficient criterion for being hired; one also had to fit into the "HP Way (the name for their culture)." Leaders in HCHP firms understand that, except for specialists, coaching and training can reliably develop ability, while the attitude and values that make for the combination of high commitment and high performance are more difficult to develop.

HCHP firms must therefore be very choosy. It is detrimental to them to hire people just to see if they work out and then dismiss the ones who don't. Southwest Airlines has historically selected only one out of 100 recruits. Some firms improve their selection ratio is by recruiting from alternative labor markets that their competitors are not likely to pay attention to; for example, social workers may be a good fit for jobs in a service industry.

The interview process in HCHP firms focuses on who the person is more than on what is listed on his or her resume. These firms typically rely on multiple interviews conducted by experienced employees, often by managers several levels above the position the company is trying to fill. This makes sense because these are the people who really know what it means to fit in that culture and who can often tell when someone won't. In an HCHP hotel I studied, the general manager insisted on interviewing *all* prospective employees. Interviewers may ask candidates to describe how they would handle certain situations, or they may ask the candidate directly about his or her values. A valuable line of inquiry is experiences that indicate something about the candidate's motivation, capacity to learn, and inclination for team versus individual accomplishment. An interviewer might ask, "Do you have a personal mission statement? Or "what is your purpose in life." If you don't, what would it be if you were to write it today?" [53]. Even when the company decides to hire someone who looks like a good fit, they do not grant employment security until that employee has proven himself or herself to fit over the course of a trial period. At Southwest Airlines it is two years. Others do not hire until the prospective employee has been through a training program that allows the company to assess cultural fit.

The criteria for such a policy of selective hiring–especially the hiring of professional and management employees cannot be copied–it must be designed by the senior team itself. That policy has to be aligned with the leadership team's strategy and plans for growth. The senior team must agree on the personal qualities that fit their strategy and values and on the personal qualities that they want to avoid. (These are not necessarily "bad" qualities that any company would want to avoid. They are qualities that may well be aligned with some other company's goals and values, but not *this* company's goals and values.) Potential is far more important in selection than capacity to fill an immediate position. I am continually amazed at how many companies look for job-specific experience and knowledge and pass up individuals who have personal qualities that indicate high potential.

Importantly, senior teams will have to keep in mind whether the firm's historic growth rate, assuming it continues, will make it too hard to be as choosy at it needs to be to become or remain a high-commitment, high-performance company. Southwest and Hewlett Packard were able to maintain the "Southwest Way" and the "HP Way" only by making sure not to grow so quickly that they could not find, hire, and socialize new people at the same pace, an example of how strategy and HR polices must be aligned with each other if high commitment and performance is to be achieved.

### 1.2.3 Socialize People to Fit the Culture

Identifying with a company is a profoundly personal and emotional process. Each individual employee must find ways to say to him or herself, "I want to fit in." Just as important, some employees must decide that "I do not fit" and leave. This occurs naturally in all organizations, but if top management wants to develop identity with and commitment to its values, they must create emotionally powerful occasions that transmit the values and norms they want to instill–occasions that will, over time, make it clear to those who fit that they do and to those who don't that they don't. Again, Jenny Chatman's research shows that those who receive the most intense socialization fit the firm's values better than those who do not [54].

There is considerable evidence from the field of anthropology, notably Joseph Campbell's research and writing, that culture is created and transmitted through myths. Campbell showed that mythological archetypes are anchored in each person's unconscious yet shared by all people in the community. It is through myths that people give up their individuality and become capable of experiencing community.[9] HCHP companies must find ways to transmit myths that exemplify their values and norms. This can occur through carefully constructed orientations for new employees, management training, company meetings, social events, and celebrations (often for long-time employees).

Designating heroes is one way to make the company's norms and values concrete and emotionally relevant. Employees can identify with real people

---

[9]For a discussion see Jarnagin, C. and Slocum, J. W. Jr. (2007) "Creating Corporate Cultures Through Mythopoetic Leadership," *Organizational Dynamics,* Vol. 36, No. 3, pp 288–302.

whose qualities and success they would like to emulate so as to one day be looked up to themselves. When I visited the headquarters of the UK grocery chain Asda (whose CEO was building a high commitment culture) to learn about its amazing transformation, there was a "wall of heroes," with pictures of employees who had achieved results or behaved in a way that reflected the culture Asda was developing. These heroes were also celebrated in company newsletters and meetings [55].

Storytelling is another way to convey myth and culture; they convey an understanding of goals and values that is logical, intuitive, and emotional" [56]. On my first case-writing visit to Hewlett Packard in 1980, I learned about several stories that were being told at orientation sessions for new employees, various management training programs, and social events. These exemplified HP's character at that time: innovation and achievement, quality, common sacrifice. There was the story about How Bill Hewlett challenged HP Labs to build a scientific calculator he could put in his shirt pocket, which led to the world's first small scientific calculator and one of HP's most important business segments. There was the story of how Dave Packard smashed up an instrument in a laboratory because he thought it was poorly designed, unreliable, and generally a "hunk of junk." And then there was the tale of the 1970 business downturn when electronics companies across the United States were laying off employees. At HP, however, every employee took a 10% pay cut and worked 9 out of 10 days–and they all kept their jobs.

Rituals are another means of transmitting a firm's underlying values and reinforcing its myths. In its heyday, Hewlett Packard held annual employee picnics at which top management did the cooking and served employees, a demonstration of the company's egalitarian culture. The ritual used at Mary Kay, a company that relies on the internal motivation of its geographically distributed female sales force, offers an especially elaborate example:

> [G]old and diamond pins, fur stoles, and the use of pink Cadillacs (Buicks in China) are presented to sales people who achieve their sales quota. . . . [A]ll participants sing the Mary Kay song, "I've got that Mary Kay enthusiasm," which was written by a sales person to the tune of "I've Got That Old Time Religion." . . . The ceremonies are reminiscent of a Miss America pageant, with all sales people dressed in glamorous evening clothes. The setting is typically an auditorium in front of a large, cheering audience. During the ceremony, when Mary Kay was introduced, she would levitate on billows of smoke to the stage. The illusion of her raising and being

kept in the air with little physical support symbolized how women could rise up and enrich their lives. During the ceremonies, bumble bee-shaped diamond pins are given to women who reached certain sales levels. The pin presents a myth that bumblebees should not be able to fly because of their aerodynamics. However, with their will power and self-determination, they can fly. The use of the bumble-bee reflects Mark Kay Ash's vision for women that with help and encouragement everyone can find their wings and fly" [56].

Engaging experienced leaders as teachers is yet another way to convey the culture and socialize employees. Because those leaders personally embody company values, they convey them much more effectively than HR managers or professional educators. When I worked with medical technology manu-facturer Becton Dickinson in the late 1980s and through the early 1990s, the company prided itself on bringing in professors from prestigious business schools–I was one of them. Today, though, Becton Dickinson has its own cor-porate university in which experienced managers–not academic professors–transmit company values by teaching how managers are expected to carry out human resource management practices such as coaching, selection, ethics, performance excellence management, and managing a diverse workforce. This and many other changes in their human resources policies and practice have enabled them to become a high commitment and performance company, something they were not when I first came to know them [58].

### 1.2.4 Develop Talent to Fit the Strategy and Culture

Leadership and management development would be a competitive advantage for any company. What distinguishes HCHP companies is how hard they work to ensure it. In contrast, a study of organizations that experienced difficulty in implementing their strategy employees identified inadequate management development as one of six barriers [59]. Many would agree with Goldman Sachs's former CEO, Henry Paulson:

> Our people have driven Goldman Sachs' success for 130 years through sustained, superb execution across a range of markets and products. The best way to maintain that advantage is by recruiting, training, and mentoring people as we always have–one at a time, with great care. We want Goldman Sachs to be a magnet for the very best people in the world–from new graduates to senior hires. At the same time, we are focusing on developing our very

deep bench of talented people and improving and extending our skills. We are, for instance, placing young leaders in demanding positions that stretch their abilities. We are also devoting more time and attention to the formal training and development of leaders, particularly senior leaders [60].

There is ample research evidence that employees' value personal development more than their paycheck, and many of us know this firsthand. Research also finds that seminal experiences lead to attachment and commitment. Moreover, setting high expectations and providing stretch assignments enables personal and leadership development. Jim McNerney, past CEO of Boeing and former senior executive at GE, described his assignment to be GE's Asian leader as a seminal experience: "[Then-CEO] Jack Welch gave me no blueprint, just said Asia is the biggest opportunity we've got, and we are not doing much–go figure it out."[10]

A shortage of leaders prevents companies from responding to new threats and opportunities–hardly a contribution to sustainability. There simply are not enough effective and culturally aligned leaders to lead initiatives, emerging businesses, or acquisitions. For every effective and aligned leader a company requires, it must also develop several managers with the potential to succeed that leader should he or she leave.

In the 1990s, Becton Dickinson concluded that it needed to grow internationally. But that meant it needed leaders who could work in teams across boundaries and who were committed to the mission of the company more than to their own country organization. They also needed a pool of leaders to draw on to lead the new businesses they knew would arise as the company grew. But these leaders were simply not available because management had selected managers for their skills to fill their immediate job rather than for their potential or fit with HCHP values which had not yet been developed. The company had also not focused attention and resources on the development task. It was a change in these policies, led by three successive CEOs, that made possible the dramatic changes in commitment and performance the company achieved by 2010.

Managers in HCHP companies like McKinsey spend a lot of time inspiring people to achieve by mentoring, coaching, and reviewing and by planning challenging assignments to broaden and develop their high-potential people. Jack Welch spent better than 50% of his time on people. Archie Norman,

---

[10]Colvin, G (2006) "How One CEO Learned to Fly," *Fortune,* October 16, 2006. [Is there an author for this, so we can have an ordinary citation?]

CEO of Asda, told me that he spent 75% of his time on human resource issues [61]. CEO Dan DiMicco, CEO of Nucor Steel, believes that, "Every person is capable of endless growth on a trail that leads we don't know where. If we'd put the holds on Cro-Magnon man, where would we be today?"[11] Boeing's CEO, Jim McNerney, says: "I don't start with the company's strategy or products. I start with people's growth because I believe that if the people who are running and participating in a company grow, then the company's growth will in many respects take care of itself. I have this idea in mind that all of us get 15% better every year."[12] Of course, such convictions can only be justified if they are translated into concrete policies and practices.

Most HCHP companies have *a policy of promoting primarily from within.* This enables them to offer developmental assignments to high potentials and it ensures that those who are promoted reflect the company's values. At McKinsey, what is needed are analytic skills, the capacity to build client relationships, and the willingness and ability to develop teamwork between talented people across the globe in support of a client engagement. At Southwest Airlines, what is needed are the friendly and cooperative attitude that is essential for good internal coordination (the kind that lets you turns planes around in 15 minutes) and a natural desire and ability to make strangers–the passengers–feel like friends. Of course a policy of promoting from within will have more difficulty co-existing with a strategy geared to very rapid growth.

Of course, an internal promotion policy can present challenges when new technology, markets, and administrative practices require skills that have not been highly valued in the firm before. HCHP firms handle that by hiring outsiders, but as few as can be gotten away with, in order to avoid diluting the culture and to minimize the risk of outsiders proving to be ineffective because they do not fit in. The reality of rapid change in technology and markets makes adhering to an internal promotion policy more difficult. Firms that want build a sustainable company must therefore, make investments in better integrating a more diverse work force into their corporate culture.

*Career paths in high-commitment companies move high-potential people across functional and business unit boundaries*, often laterally. This really puts a person's ability to motivate and lead others to the test because one cannot rely on one's previously acquired functional knowledge to solve problems. At Hewlett Packard, this process was known as the "career maze," reflecting its lack of predictability. In contrast, research by Russell Eisenstat,

---

[11]Nucor Steel's Culture (draft) with permission © Helen Kelly/The Working Manager Ltd 2006/2007.

[12]Colvin, G. (2006). "How One CEO Learned to Fly," *Fortune,* October 30, 2006.

Bert Spector, and I found that in companies that were not competing success-fully, like Bethlehem Steel in the 1980s, senior executives had risen vertically through one function (Beer, Eisenstat, & Spector, 1990). Successful cross-boundary experiences develop a manager's leadership skills by cultivating the general management perspective that is so important for business leadership.

*High-performance companies evaluate their people, particularly their managers, differently than other companies differently.* People are evaluated not only on the results they achieve but also–just as consequentially–on their fit with the company's values and culture. People who behave in ways that violate cultural norms are weeded out (Figure 1.3). To create a team-based culture at Morgan Stanley, CEO John Mack introduced 360-degree evaluations–this is, evaluation by one's boss, one's peers, and oneself, something that more and more companies are emulating. Even a successful investment professional would be in trouble, in terms of pay and promotion, if peers judged him or her to be selfish, uncooperative, or overcompetitive. This was a radical step in the investment banking industry, where bringing in the big clients and the big bucks made up for just about anything. When Mack heard that senior leaders in London had tried to take their people to another firm, clearly demonstrating that they were not committed to Morgan Stanley, he personally flew to London on an overnight flight, fired the uncommitted leaders, and made it public in subsequent speeches to employees.

*High-commitment, high-performance companies avoid filling top positions with outsiders.* Precisely because such outsiders have already had a long

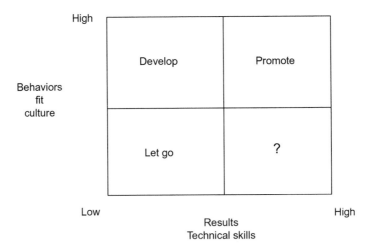

**Figure 1.3**   Criteria for promotion in high commitment and performance companies.

history of success elsewhere, there is a good chance of them not fitting the company's values and culture and therefore failing in their new job. This is particularly true with regard to CEO succession. Outside CEOs have difficulty understanding the culture emotionally and therefore sustaining it. Because HCHP companies maintain a strong bench, they can generally avoid this risk; they can grow without finding that they've run out of managers who embody the culture of the firm.

The importance of human resource management practices such as career paths, goals, and evaluation for the development of high-potential leaders is borne out by seminal research conducted in the 1980s by Morgan McCall and his associates at the Center for Creative Leadership. They found that, on average, an individual manager's derailment was as much a function of the company's human resource practices as it was a function of that manager's personal characteristics [62]. Intelligent and technically efficient managers derailed when the company's policy was to move managers rapidly up through narrow vertical functional channels. Managers with impressive results and strong track records derailed if the company did not evaluate and consider personal flaws when making development and promotion decisions; that is, when the company left flaws in place until those flaws finally proved fatal. Managers who were intelligent and achieved great results also derailed if the company focused entirely on short-term objectives. These findings explain why human resource practices such as cross-boundary career paths, long-term goals, and evaluation of interpersonal capabilities as well as of results enable HCHP firms to develop a deep bench of leaders who are not only skilled but are also dedicated to their *collective* success. These findings also make clear that a high-commitment, high-performance HR system depends heavily on the top management team's commitment to a practice they must lead, not just on the HR Department.

*To succeed, leadership development must be motivated and managed by the corporate center and must involve senior teams at every level.* The first task is for senior teams to define the essential capabilities needed to carry out their business model and the values they want to guide behavior–by stimulating some behaviors and limiting others. The HR function can then develop the necessary practices in detail and can help manager implement them.

At Becton Dickinson, for example, three successive CEOs and their senior teams transformed its selection, promotion, and development policies and practices so that these would all be aligned with the company's strategy and values. In 1990, CEO Gilmartin and his senior team, whose management review committee had previously met only irregularly, sat down to define essential capabilities and values. Each senior executive identified the managers who, in their view, fit the strategy and the high-commitment values they had defined. Having discussed and ranked these managers, the senior team created a list of personal qualities they considered critical to improving the company's effectiveness, performance, and commitment. Those criteria and the practices defined in 1990 have been redefined over the years, both as circumstances changed and as the company learned from experience. Again, it the responsibility of leadership teams to lead this learning process with HR in support.

Becton Dickinson also developed an elaborate system of bottom-up review meetings that culminated at the very top. CEO Ludwig and his team then used these bottoms up reviews as well as their observations to identify and then develop promising managers who might succeed to top management jobs. Business unit managers are also required to have a minimum number of replacements for themselves. To ensure that this process is carried out at all levels, BD does not pay out a bonus unless the evaluations and development plans have been submitted. At GE, review of key managers in each business were attended personally by Jack Welch; this was his opportunity to get a feel for the best people and ensure that promotions matched corporate strategy and values.

Each HCHP company will have its own developmental practices, but HCHP companies all make time for key managers to discuss and develop people. The process itself has a subtle benefit: it reminds the managers involved of the company's performance standards and values and has the potential to recommit them to those high standards. At the Harvard Business School, for example, which has been an HCHP organization for 100 years, I can attest that the reinforcement of standards and values is particularly powerful. All senior faculty (around 90) read each promotion "case" (a 20-page report written by a committee about the candidate's achievements) and meet together to discuss the person's qualifications and then vote on the promotion. I don't think many faculty members could sit through this process without being reminded of what the school standards are, without evaluating their own accomplishments in that light, and without being at least a little humbled and recommitting themselves to high performance.

Although stretch assignments and boundary-less career paths are the primary means HCHP firms use for developing managers, they offer extensive education and training to supplement on-the-job development. Research has shown, however, that these can easily become the "great training robbery" (meaning that there is little return on investments) if senior leaders do no embed them in their intentional effort to transform the company's system of organizing, managing and leading [63]. The best educational experiences include not only classroom work but also assignments to assess a business or an organizational problem and recommend solutions to senior management. When experienced managers teach and coach in these programs, as they do at Becton Dickinson, the firm's culture is transmitted and reinforced by those who deliver the training.

## 1.2.5 Rewards

HCHP companies provide extrinsic rewards, such as money and status, but they are primarily organized and managed to provide higher intrinsic rewards such as challenging and meaningful work, self-actualization, and participation in a community of purpose. When you have employees who are more internally rather than externally controlled, you have commitment. F. Kenneth, Iverson, former CEO of Nucor, believed that if you give people responsibility and get out of their way, they will achieve extraordinary results. He turned the company around by redesigning it to give mill managers full responsibility for achieving results with as little interference from corporate as possible. Likewise, turnarounds at Hewlett Packard, AES Corporation, Asda, and Southwest Airlines were accomplished by shifting control from the corporate center to smaller units closer to the actual value-creating work–such as Asda's stores and Southwest's airport crews–and involving them in decisions. Again, another example of how leader's management philosophy is essential in shaping human resources policies and practices.

Many studies (too many to cite here) have shown that intrinsic rewards have more influence over behavior than extrinsic rewards do. To take just one example, a "Quality of Employment Survey" that found:[13]

- 76.3% disagreed with the statement, "I'd be happier if I didn't have to work at all."
- 72% responded that they would continue to work even if they were financially set for life.
- Approximately half agreed with the statement, "What I do at work is more important to me than the money I earn."
- More than half said they disagreed with the statement, "I work to get money in order to do other things."

---

[13]The Quality of Employment Survey is conducted by the National Business Research Institute, a leading business research and consulting firm committed to achieving their full potential.

- 58% reported that they put in "*a lot* of extra physical and mental effort into their jobs beyond that which is required."

HCHP compensation systems are certainly not all the same, but they have a common goal: to compensate people well–usually above the industry average–without undermining the work itself and the meaning of that work as motivators. It is a tricky balance: they want to pay people well without creating an environment in which people are working (or applying for jobs there) primarily for their pay. They achieve this balance by making their challenging, high-commitment, high-performance climate–*not* compensation–the primary basis for recruiting and motivating employees. Note that implementing such a policy has to begin with the executive team defining the necessary intrinsic motivations; only then can the HR department implement the appropriate procedures and policies.

Although this philosophy is intuitively obvious to HCHP leaders, it is supported by research which finds that a focus on money can actually *undermine* intrinsic motivation. People come to see money as the reason for their effort, which in turn reduces the importance of the meaning of one's work and of one's commitment to the company's mission and purpose [64]. HCHP companies therefore avoid tying compensation too tightly to *individual* goal accomplishment, but rather to the collective accomplishment of the team and of the company as a whole. In fact, this is not only idealistic but realistic: one's performance in a job is typically so dependent of the performance of others that it is almost impossible to accurately measure an individual's efforts and abilities. It is very seldom that a job is so simple and the circumstances are so stable than an individual's performance outcomes aren't affected by anyone else's performance.

For several reasons, HCHP companies are wary of any kind of pay-for-performance system, often avoiding them altogether. Research by Beer and Katz shows that tightly coupling pay to an individual's results encourages narrow, self-serving behavior, often in the form of "gaming the system" [65]. It also encourages distorted comparisons and competition, which in turn lead people to feel they are not being recognized or treated fairly. Of course, that undermines trust and commitment, but it also undermines group performance because managers have to spend fundamentally unproductive time administering and redesigning pay systems to deal with complaints. Yet another problem is that a sense of control over one's own work–one of the foundations of commitment and pay-for-performance systems–inherently convey to people that management is in control. Finally, recall the results of

Morgan McCall's study, mentioned earlier, showing how pay-for-performance contributed to the derailment of high-potential managers [66].

HCHP leaders treat incentive systems as, at best, a necessary evil. Designing one is therefore an exercise in damage control: the goal is to do the least harm possible to motivation and performance. John Clarkson, former CEO of the Boston Consulting Group, explained that he views his compensation system as an exercise in justice: design your compensation system to achieve fair outcomes and use intrinsic means, such as culture, mission, and challenging work, to motivate the desired behaviors. Skepticism about the motivational power of money to motivate performance is supported by a body of research too extensive to cite here.[14] In one noteworthy study, Alan Binder, a Princeton economist and former Vice Chairman of the Federal Reserve Board, reviewed many studies on the subject of pay and motivation of lower-level workers (not managers) and found, to his surprise, that how employees were treated and how much they were engaged in work had a greater effect on productivity than the system of pay for individual performance did. The most effective systems, he found, were stock ownership by all employees and profit- or gains-sharing plans combined with worker participation. These systems use rewards not to motivate but rather to enable employees to feel that they are part of a community engaged in a collective effort to achieve the organization's mission and results [67].

My own research and that of others suggests that Binder's counterintuitive findings are true of senior executives as well. While teaching in Harvard Business School's Advanced Management Program, I found the executives in my class disagreeing with my skeptical of view of pay-for-performance systems; they believed that such systems definitely did more good than harm. (US executives, in particular, were believers; European and Japanese executives not quite as much.) So, I asked them to participate in a survey study of their companies' compensation system and its effects. I found *no* relationship between the percent of total compensation (bonuses and stock options) at risk if performance targets were not met and (a) the reported motivation of managers in the firm, (b) the effectiveness of the firm's organization, and (c) the firm's actual financial performance [69]. In fact, I found–just as Binder

---

[14]For much more detailed discussion of this important subject see Pfeffer, J. (1988) "Six Myths About Pay," Harvard Business Review May-June; Pfeffer, J., (1998) The *Human Equation: Building profits by putting people first,* Boston, MA, Harvard Business School Press; Milkovich, G. and Wigdore A. (1991) *Pay for performance: Evaluating Performance Appraisal and Merit Pay,* Washington D.C., National research Council; Ehrenberg, R.G. (ed) *Do compensation policies matter?* Ithaca, NY, Cornell ILR Press.

had–that the most effective reward systems were ones that tied pay to the firm's overall performance. Most interesting, the only factor in this study that predicted firm performance was the extent to which executives reported a *participative and team-based culture.*

Not surprisingly, I found that executives of firms with a higher percentage of pay at risk reported more gaming behaviors and dysfunctional outcomes such as setting low targets, lack of cooperation with other units, and an unacceptably high percentage of time spent administering and redesigning compensation systems in response to executive complaints. In my experience and in my research, I find that any executive compensation system that does not pay off as expected is seen as a bad system, leading to high administrative and redesign costs and less focus on fixing real problems and performance [69, 70]. Such systems were also associated with much larger ratios of CEO pay to that of the lowest-level employee (500-to-1 compared to ratios as low as 40-to-1), making it much harder to create an egalitarian culture and a mission-driven community of purpose; this has also been well documented in other studies [71]. Again, this means that top managers of HCHP companies have to have values that enable them to accept lower pay than their peers.

Reviewing a large body of literature, Baron and Kreps, conclude that "[pay for performance schemes are too blunt an instrument, resulting in misalignment of incentives; they load uneconomically large amounts of risk on the worker; they often have problems with perceived legitimacy; they can breed inflexibility; and they can dull intrinsic motivation" [72].

Why then do so many companies pursue compensation policies that undermine effectiveness, commitment, and performance? My own Harvard study gives us some idea: the executives in my class were themselves not entirely sure that compensation motivates, but what they were sure about is that *their* bosses–top management–believed it did [73]. Although the executives in my class were well aware of the dysfunctional effects of incentive programs, the top managements of their companies seemed not to know or else not to care. We saw how that kind of thing played out in the 2008 financial meltdown; the promotion of subprime-mortgage-backed securities was propelled by intense incentive systems–formal and informal–with little or no regard to the consequences of the behavior that was being so powerfully incentivized. When a task force of large business asked to report on barriers to the organization's effectiveness reported that sales incentives were undermining effectiveness, senior executives refused to eliminate them. The same mistake was repeated spectacularly at Wells Fargo, where a sales incentive led to unethical selling practices first widely publicized in 2016 and led to huge penalties and ultimately the removal of the CEO and many board members. There is no end of other examples.

Are incentives absolutely to be avoided? Could a firm that already had a strong high-commitment, high-performance culture improve its commitment and performance even further with incentive systems contingent on financial results? Surely two powerful levers are better than one. But remember that one of the pillars of HCHP is the *performance alignment* of all structures, systems, people, and culture. I have found, however, that incentive systems introduced into a high-commitment culture undermine that culture because they are not aligned with the rest of the culture, which emphasizes teamwork and dedication to the company's goals rather than to specific individual goals. Here's an example: Under competitive pressures in the late 1980s and early 1990s, a number of business-unit general managers at Hewlett Packard convinced senior management to let them introduce pay-for-performance systems, which had never been allowed there before. Within three years, the very same executives who had lobbied for a pay-for-performance decided to stop it [74] Employees saw unfairness in the system, which of course undermined trust. Managers found that the pay-for-performance incentives shifted people's attention–including their own–away from continual improvement of their unit's performance and culture building. The problem is that pay-for-performance creates a direct line from individual action to pay and thus undermines dedication to a community of purpose–which is where all the good stuff happens in terms of team work and company performance.

Collaboration between Stanford researchers and McKinsey confirms the idea that pay systems must fit cultural objectives [75]. The study found that firms using "soft" motivation levers–relationships and culture–did not use levers such as incentive systems. The only exception in their sample was Enron, where major ethical lapses caused the demise of the company–enough said. Both this study and the Hewlett Packard experience suggest that HCHP companies that adopt conventional pay-for-results incentive systems risk poisoning their cultures with self-serving, dysfunctional, and even unethical and/or illegal behavior.

Executives sometimes insist that they have no choice but to use pay and incentives to recruit and keep talent. If they do not do it, they will be consistently outbid for the best people. New recruits cannot be offered lower pay than the labor market dictates. But they also do not have to be offered outsized wages and bonuses to attract them to a high commitment company. HCHP companies are looking for a different kind of "best"–the best fit with their HCHP high involvemnet cultures. Their own particular kind of "best". Their compensation principles include the following:

**Use non-monetary rewards to motivate.** HCHP firms emphasize rewards such as more responsibility and freedom, challenging work, private and public recognition, symbolic awards, and fun. Compensation is seen as a recognition that you have done good work (and that you are a person with a life that requires you to earn a living) but not as a prior inducement to do it.

**Ensure that *total compensation* is well above the median.** Although HCHP companies emphasize fit, they do need skills and knowledge, which they won't be able to attract or retain without good pay. Costco, an HCHP firm, pays its employees approximately 40% above the market and still manages to achieve extraordinary market performance. Senior management frequently has to reject the advice–or demands–of Wall Street analysts to reduce wages to industry-average levels in the name of shareholder value. Wall Street does not appear to understand the difference between cost and productivity per employee; it is high productivity, not low cost, that gets you long-term high performance and resilience.

**Practice symbolic egalitarianism.** HCHP firms generally avoid the trappings of hierarchy. They are not fond of organization charts and well-defined career and salary ladders. What they do is to create very broad pay ranges intended to deemphasize hierarchy and instead encourage employees to positions in which they are needed and–very important–in which they will develop. (Remember, HCHP firms see their people as *investments*.) Everyone, from senior management to the lowest levels, is generally subject to the same compensation and benefit plans and distinctions are minimized with regard to perks such as offices, titles, and healthcare policies. At Toyota, for example, the performance bonus–based on overall business performance, not individual performance–is a percent of your wages, but it's the same percent for everyone. Soon after becoming CEO of Asda, Archie Norman abolished all offices, including his own. HCHP companies that do think private offices are necessary make sure that all offices are the pretty much the same.

To leaders who mean to develop a community of purpose, it seems obvious to organize the company this way. But these practices are also supported by research in cognitive psychology, including survey research conducted in actual workplaces. Such research finds that people tend to use cues around them to make sense and form judgments of their social experience. HR policies and practices and the expectations communicated by others, particularly those in high-status positions are among the cues. For example, firms with "more generous" employment policies and well-defined career ladders have been found to have high employee dissatisfaction" [76].

Symbols of hierarchy and status are also unwanted by executive teams who want upward feedback to let them know when something is wrong while there is still time to fix it. Employees use such symbols, or their absence, to gauge how safe it would be to present their bosses with some bad news.

**Practice "safe" pay-for-performance.** We already discussed the threat that tightly coupling pay for individual performance systems pose to a high-commitment, high-performance culture. Nevertheless, even HCHP firms must in some way recognize the differences in performance between one person and another if the total system of compensation, rewards, and recognition is going be seen as fair. One way out of this trap is to make compensation contingent on multiple factors: achievement of individual goals, business-unit performance (often based on a rolling average of several years), corporate performance, and how consistent the individual's behavior has been with the firm's goals and values. Lincoln Electric, for example, was able for many years to use a piece-rate incentive system that constituted 100% of production workers' pay. They did it by making those incremental payouts contingent on collaboration with others. Without this additional measure, such an incentive would guarantee of all kinds of cheating and gaming and would certainly deter cooperation.

Besides basing compensation on a blend of hard and soft performance outcomes, HCHP firms may also base it on the judgment of several managers–not just one's own immediate boss–in order to gauge the extent to which one takes other teams and units into account. This, in turn, requires some training for the managers doing the appraising and requires feedback and tinkering to keep the system fair as circumstances change.

**Create *formal mechanisms* by which employees can voice concerns about compensation.** To ensure that compensations are fair and are seen as fair, some HCHP firms involve employees in the redesign of the pay system. Cummins's engine manufacturing plant in Jamestown, New York, did this in the 1970s. Although management was at first worried that employees would make greedy demands, they found the proposals from a team of employees to be reasonable and effective [77]. Ed Lawler, who has studied compensation extensively, has also found that employee involvement in the design of pay systems can work out well [78]. These findings are consistent with the literature on what academics call "procedural justice [79]". People are more likely to feel that the outcomes of a decision process (for example, a trial) are fair if they understand the decision-making process itself and feel that it is inherently fair. That, in turn, is more likely if they themselves have had a say in creating or carrying out that process.

**Keep senior executive pay in line.** Of all the possible symbols of company hierarchy, probably the most potent is how much top executives–especially the CEO–are paid. Paying the CEO hundreds of times what ordinary workers earn is likely to be seen, felt, and resented as grossly unfair, no matter what objective arguments are brought to bear. Most senior executives in HCHP firms, though they are well paid, earn less and get fewer and smaller perks than their peers at other companies and thus their compensation is not as dramatically out of proportion to that of ordinary workers in their firms. A 2006 study of 2,275 US companies, for example, found that the CEOs of the 25 worst-performing (in terms of shareholder wealth) had an average pay package of $16.7 million while the CEOs of the 25 best-performing were paid $ 4.4 million on average [80]. So much for the argument we keep hearing that companies have no choice but to overpay their CEOs or they won't be able to hire the best.

### 1.2.6 Managing the Inevitable Crisis

A company doing its best to achieve high commitment and high performance will nevertheless one day face a crisis in performance. How top management handles that crisis will determine whether the organization eventually moves ahead on its path toward high commitment and high performance or squanders its investment in the culture and the talented people it has spent years developing. It often seems that the only way to deal with a crisis is to do something that knocks down one or more of the pillars of HCHP–"emergency measures" such as layoffs, cutting training from the budget, or lowering quality or service to cut costs. It may not be impossible to recover the trust and commitment that are lost this way, but it's darned near impossible. The companies that avoid this trap are the one whose understanding of HCHP shows them, first to avoid risky strategies of rapid growth, too many rapid-fire acquisitions and the high debt that is typically required to peruse these policies. They also find some other way to survive the crisis–a way that is not fundamentally unaligned with HCHP principles. And at such a crisis, the company's HR system–how it has selected, trained, socialized, promoted, and rewarded its people all along–will be decisive.

In June 1988, for example, Dreyer's Grand Ice Cream, a $1 billion company that had been doing quite well, found itself in a world of trouble [81]. An expansion project was off schedule and over budget, so the expected profits weren't coming in yet. Meanwhile, the price of butterfat, the key ingredient in ice cream, rose to a record level–just what they needed. Dreyer's could not get out of this jam by raising its prices, even temporarily, due to aggressive discounting by its chief competitor. Revenue in one of Dreyer's new product lines began to drop and one of its customers was threatening to terminate a long-term distribution contract.

Of course, investors and analysts wanted and expected Dreyer's to start restructuring and cutting costs immediately. But senior management, including Gary Rogers and Richard Cronk, had spent many years building an open, team-based culture and in establishing their own accessibility and commitment to their employees. They were not going to throw that away now; instead, they were going to rely on it. When they were prepared to announce their restructuring to the financial community, executive committee members were on airplanes to talk with every one of their 400 employees. Cronk observed, "We know our limits and understand the law, but we tend to be very open with our employees, we communicate a lot." As one account executive recalled, "They reassured us by calling it straight... They informed us of their game plan and that they needed us... You looked at these [senior managers] and thought, you'd run through a wall for this guy [82]".

Dreyer set up an 800 number so employees could call up and hear CEO Rogers's pre-recorded speech about the situation and his plans. The speech was honest and owned up to the company's big problems, but also was upbeat. Senior management continued to invest in the Dreyer Leadership University (DLU), making it clear that employee development was the top priority they had always said it was. Nor was this just a symbolic gesture; Cronk and others in senior management considered it an investment in the future that would pay off. "When people heard that we were investing another million dollars into the [culture] and DLU it created a high degree of comfort and confidence that we're focused on what really matters," observed the VP of Sales [83]. For one thing, it proved that top management still thought there *was* a future.

The company's revenue, profits, and stock price all came back. By 2001, the stock had risen from 9.88 in 1988, the year of the crisis, to 36. By January 2003, it had risen to 71.23. Reflecting on how Dreyer had handled the crisis, Cronk said, "It was a common trust and of sharing the facts–openness.... We were not sugarcoating anything, putting a Hollywood spin on anything....

We were honest and clear. . . People believed the story and they understood. . . There was an enormous amount of pride and optimism [84].

Even in the most well-handled crises, however, there is bound to be loss and damage. Some people will lose their positions or even their employment. This loss and damage must be dealt with in a way that leaves the three pillars intact and still credible. And as any first responder can tell you, an emergency will be handled best by a team of people who have planned and practiced for emergencies together. HCHP companies therefore need to develop a set of policies in advance of any crisis that will minimize damage from unavoidable restructuring and downsizing and will maintain employee dignity and commitment. Wayne Cascio, who has studied companies with a record of stable long-term employment contracts, lists the following policies to preserve the dignity of both those who have to leave and those who get to stay and to preserve the company's relationships with both groups [85] "Use downsizing as a last resort; at the same time, reinvent your business." There are several ways companies have done this:

o Rely on attrition to right size-cut your employment to required levels.
o Use redeployment and leave layoffs as an absolute last resort. Hewlett Packard used this strategy extensively throughout the 1980s as it faced intensified competition. Employees were given three months to find a job in HP and helped to do so. If they could not find an equivalent-level job, they were offered a lower-level job. If they did not want it, they were given generous severance packages.
o Ask for volunteers who want to take extended vacations, a sabbatical, a leave of absence, or a shorter work week.
o Ask everyone to share the pain by taking a pay cut. Senior management should take an even larger one. Hewlett Packard used this approach several times in its early history.
o Shorten work weeks and offer stock options in return.
o Lend employees to nonprofit firms and pay them the difference in their wages.
   • "Do everything you can to manage survivors well." Tell survivors exactly how departed employees are being treated.
   • "Generate goodwill, even loyalty, among departing employees" by taking great care with how they are separated. Generous severance packages, outplacement services, and retraining are typical strategies.

There are three principals involved here. First, HCHP companies do as well as they possibly can by anyone they have to lay off or transfer to a less-desirable job simply because that is how they feel people should be treated. Second, they are aware that the employees who stay are watching how the others are treated. Those are their friends and former co-workers–and, down the road, it could be *them*. Third, HCHP companies realize that if business picks up, they would much rather rehire their former well-trained, well-acclimatized employees than new ones, so it will pay not to make them so angry they would never come back.

An HCHP company with a profit-sharing plan may find it easier to retain its employees during a crisis. In Japanese companies, for example, 25% of employee pay is based on profit-sharing. If profits suddenly drop, so does the labor portion of the company's cost structure, which is not any fun for anyone but does allow the company to avoid layoffs.

## 1.2.7 Unions

While only 10–11% of the US workforce is unionized, the percentage is much larger in Europe and Asia. A unionized workforce can be an obvious obstacle to establishing a high-commitment, high-performance culture; a union pre-supposes an "us-and-them" reality that is opposed to the non-hierarchical, organization-centered spirit of HCHP. Nevertheless, HCHP practices have been implemented in unionized settings, though not all high-commitment, high-performance HR policies and practices can be implemented. Some of the earliest experiments in high-commitment manufacturing plants took place at Proctor and Gamble and General Motors, both of which were unionized. The key in all of these experiments was to develop a partnership between the union and management. That means recognizing the union's legitimacy (that is, its right to represent these particular workers), involving it in strategy discussion and plans, respecting voice and governance mechanisms, and involving workers at all levels in solving problems.

Southwest Airlines, for example, is unionized and has managed sustained high performance–not by finding a way around the unions but by seeing them and treating them as partners. That complicates the management task, but it also provides some protections. Unions are an excellent check on management; they can prevent it from making mistakes and can help mobilize workers.

In Europe, labor-management cooperation is legislated through insti-tutional mechanisms such as works councils (they have to be consulted by management) and industrial democracy in Germany in which corporate boards must have labor and worker representation in all strategic decisions. This mandated cooperation did not deter Germany's well-documented economic success after World War II to the present. Cooperation, contrary to assumptions of most US managements, has actually proven to enhance firm adaptivity rather than reduce it [86].

## 1.3  Challenges to High-commitment, High-performance HR Systems

Despite all the benefits, high-commitment HR systems present some potential problems of which leaders need to be aware.

- A cult-like culture can cause employees to become inwardly focused unless the culture is designed to value innovation and change and their necessary precursor–a diversity of ideas. There is always a tendency for groups to select, promote, and value people like themselves. In the 1990s, Hewlett Packard's executive team realized that its strong male-dominated engineering culture had made it hard for women to progress. The CEO, Lou Platt, led a serious effort to recruit, develop, and train women. By the late 1990s, HP had one of the largest groups of women senior executives among high-technology companies.
- High-commitment HR systems are challenging to establish in industries, such as retail, hospitality, and restaurants, with high turnover. Employees in these industries tend to see their job not as a career but rather as something to pay the bills until they can find something better. Even so, a determined executive team can go a long way towards establishing high commitment and high performance. Peter Dunn, the CEO of Steak n' Shake, a restaurant chain, was able to reduce turnover by 40% by establishing sustainable human resource practices. He established a management philosophy that asked leaders at all levels to lead through high involvement of their people [87]. A seminal study of what makes for excellent service by Harvard my colleagues Sasser, Schlesinger and Heskett have shown that focusing on employee satisfaction and commitment can materially improve commitment and performance in the service sector [88].

In a seminal 2014 book "The Good Jobs Strategy," Zeynep Ton describes how retail companies succeed through counter-conventional human resource practices, investing in their people for the long run by creating meaningful jobs, paying well, providing full-time meaningful jobs in an industry famous for scheduling people at will depending on expected store traffic [89]. CEOs and their boards resist short term pressure for quarterly earnings. The CEO of Costco, a grocery and retail company, famously resisted Wall-Street analysts' pressure to cut employee wages as I noted earlier (they were some 30% above average for the industry) because he understood that attracting and keeping high-commitment employees was what made the company so successful. The CEO of Patagonia, a retail company, has followed a similar path of higher the average wages, high-involvement jobs.

- National cultures, institutions, and laws vary and, in some countries, can seem very opposed to HCHP practices. Yet companies have made their way to high commitment and high performance in a variety of cultures, indicating that the necessary systems and practices–including the HR system–can be adapted to local circumstances. Hewlett Packard, which I studied in the 1980s and 1990s was able to implement many of its egalitarian and high involvement human resource practices as it grew internationally. When I spoke about Hewlett Packard's (HP) high-commitment human resource policies and practices and why they work to a group of senior executives in the 1980s the head of the company's Mexican organization raised his hand and with a great deal of excitement. "Now I understand why the Hewlett Packard organization on the fifth floor of our office building do some crazy things. They have no private offices for the senior executives and let lower levels address them by first name." HP's commitment to their culture and supporting polices, caused them to drop their partnership with a Korean company because they resisted implementing HP's high-commitment polices.
- HCHP is way for companies to thrive in a constantly changing environment. A company in a particular industry that is either very stable or oligopolistic may have little to gain from investing in a high-commitment, high-performance culture and its attendant HR system. Employee commitment just won't make that much difference. However, there aren't many such industries left.
- Finally, sustained commitment to sustainable human resource policies and practices requires CEO's and their boards of directors to resist short-term pressures to cut costs in order to deliver quarterly earnings. The only way they can do this, is to leverage the commitment of people to create a culture of continuous improvement.

## 1.4 Conclusion

A business strategy is only as sustainable as the company carrying it out. That requires a human resources system designed to perform, develop commitment and capacity to learn, change and adapt. In this chapter, I have outlined the elements of the HR system required for what I call a high-commitment, high-performance company; that being–in my view–the only kind of company likely to be sustainable in the current, ever-changing business environment. It is also the most human centric system with an ethical and moral foundation. Such an HR system cannot be grafted onto a company by copying other high-commitment, high-performance companies. It must be rooted primarily in the CEO's and the executive team's development of a winning company's strategy, human values and their own efforts to craft human resource policies and practice consistent with their strategy, values and their particular circumstances. Human resource managers play an important role in supporting and partnering with senior management's in developing a sustainable human resource system.

## References

[1] Gittell, J. H. (2005). The Southwest way: using the power of relationships to achieve high performance (New York, McGraw Hill).

[2] Heskett, J. L. and Hallowell, R. H. (1993). Southwest Airlines (A), Harvard Business School Case No. 694-023 (Boston: Harvard Business School Publishing).

[3] O'Reilly, C. and Pfeffer, J. (1995). Southwest Airlines: using human resources for competitive advantage (A) & (B), Stanford Graduate School Case Nos. HR1A and HR1B.

[4] Birger, J. (2002). The 30 best stocks from 1972 to 2002. Money Magazine 31 (11), 88.

[5] Tully, S. (2015). Southwest's Radical New Flight Plan. Fortune 172, 5), (October 1, 2015), 138–136.

[6] O'Reilly, C. and Pfeffer, J. (1995). Southwest Airlines: using human resources for competitive advantage (A) & (B), Stanford Graduate School Case Nos. HR1A and HR1B.

[7] Heskett, J. L. and Hallowell, R. H. (1993). Southwest Airlines (A), Harvard Business School Case No. 694-023 (Boston: Harvard Business School Publishing).

[8] O'Reilly, C. and Pfeffer, J. (1995). Southwest Airlines: using human resources for competitive advantage (A) & (B), Stanford Graduate School Case Nos. HR1A and HR1B.

[9] O'Reilly, C. and Pfeffer, J. (1995). Southwest Airlines: using human resources for competitive advantage (A) & (B), Stanford Graduate School Case Nos. HR1A and HR1B.

[10] Gittell, J. H. (2005). The Southwest way: using the power of relationships to achieve high performance (New York, McGraw Hill).

[11] Barron, J. N. and Kreps, D. M. (1999). Strategic human resources: frameworks for general managers (New York: Wiley). Chapter 3.

[12] Porter, M (1996). What is strategy. Harvard Business Review. 74 (6) 61–78.

[13] Khurana, R. (2007) Searching for a corporate savior: The irrational quest for charismatic CEOs. Princeton, NJ; Princeton University Press.

[14] O'Reilly, C. and Pfeffer, J. (1995). Southwest Airlines: using human resources for competitive advantage (A) & (B), Stanford Graduate School Case Nos. HR1A and HR1B.

[15] O'Reilly, C. and Pfeffer, J. (1995). Southwest Airlines: using human resources for competitive advantage (A) & (B), Stanford Graduate School Case Nos. HR1A and HR1B.

[16] Hornstein, H. A. (2002). The haves and the have nots: the abuse of power and privilege in the workplace and how to control it (New York: Financial Times Prentice Hall).

[17] Barron, J. N. and Hannan, M. T. (2002). Organizational blueprints for success in high-tech start-ups: lessons learned from the Stanford Project on Emerging Companies. California Management Review 44 (3), 8–36.

[18] Cameron, K. (2008). Positive leadership: strategies for extraordinary performance (San Francisco: Berrett-Kohler).

[19] O'Reilly, C. and Pfeffer, J. (1995). Southwest Airlines: using human resources for competitive advantage (A) & (B), Stanford Graduate School Case Nos. HR1A and HR1B.

[20] O'Reilly, C. and Pfeffer, J. (1995). Southwest Airlines: using human resources for competitive advantage (A) & (B), Stanford Graduate School Case Nos. HR1A and HR1B.

[21] Bauman, R. P. (1998) Five Requisites for implementing change. in Hambrick, D. C., Nadler, D. A. and Tushman, M. L. (eds.) Navigating change: how CEOs, top Teams and boards steer transformation (Boston, MA, Harvard Business School Press).

[22] Worley, C. G., Williams, T., and Lawler, E. E., III, (2014). The agility factor: building adaptable organizations for superior performance (San Francisco: Jossey-Bass).

[23] Miller, D. (1986). Configurations of strategy and structure: towards a synthesis. Strategic Management Journal 7, 233–249.

[24] Schneider, B. (2002) Climate strength: A new direction For climate research. Journal of Applied Psychology, 87 (2), 220–229

[25] Miller, D. (1987). The genesis of configuration. Academy of Management Review 12, 686–701.

[26] Gittell, J., Cameron, K., and Lim, S. (2006). Relationships, layoffs and organizational resilience: airline industry responses to September 11th. Journal of Applied Behavioral Science 42 (3), 300–329.

[27] Gittell, J., Cameron, K., and Lim, S. (2006). Relationships, layoffs and organizational resilience: airline industry responses to September 11th. Journal of Applied Behavioral Science 42 (3), 300–329.

[28] Gittell, J. H. (2003). The Southwest way: using the power of relationships to achieve high performance (New York, McGraw Hill).

[29] Barron, J. N. and Hannan, M. T. (2002). Organizational blueprints for success in high-tech start-ups: lessons learned from the Stanford Project on Emerging Companies. California Management Review 44 (3), 8–36.

[30] Pfeffer, J. (1998). The human equation: building profits by putting people first (Boston: Harvard Business School Press).

[31] Gittell, J. H. (2003). The Southwest way: using the power of relationships to achieve high performance (New York, McGraw Hill).

[32] Gittell, J. H. (2003). The Southwest way: using the power of relationships to achieve high performance (New York, McGraw Hill).

[33] Lawrence, P. and Lorsch, J. (1967). Organization and environment: managing differentiation and integration (Boston: Harvard Business School Press).

[34] Eisenhardt, K. M., Kahwajy, J. L., and Bourgeois, L. J. (1997). How management teams can have a good fight. Harvard Business Review 74 (4), 77–85.

[35] Gittell, J. H. (2003). The Southwest way: using the power of relationships to achieve high performance (New York, McGraw Hill).

[36] Beer, M. (forthcoming on 2019) Fit to Compete: Honest, collective and public conversations about your organization will transform its effectiveness. (Boston, Harvard Business Publishing).

[37] Beer, M. and Eisenstat, R. (2004) How to have an honest conversation about your strategy. Harvard Business Review 5925 (February 2004), 82 (2) 82–89.

[38] Nocera, J. (2008). The Sinatra of Southwest feels the love. The New York Times (May 24), C1 https://www.nytimes.com/2008/05/24/business/24nocera.html?mtrref=www.google.com&gwh=4D2A6109B0 049FA5E56A87BE18421735&gwt=pay

[39] O'Reilly, C. and Pfeffer, J. (1995). Southwest Airlines: using human resources for competitive advantage (A) & (B), Stanford Graduate School Case Nos. HR1A and HR1B.

[40] Barron, J. N., and Kreps, D. M. (1999). Strategic human resources: frameworks for general managers (New York: Wiley). Chapter 3.

[41] Pfeffer, J. (1998). SAS Institute (A): a different approach to incentives and people management practices in the software industry. Case no HR-6, Stanford University, 5.

[42] Pfeffer, J. & Sutton, R. I. (2006) Hard facts, dangerous half-truths, and total nonsense: Profiting from evidence-based management. (Boston MA, Harvard Business School Press.)

[43] Pfeffer, J. (1998). SAS Institute (A): a different approach to incentives and people management practices in the software industry. Case no HR-6, Stanford University, 5.

[44] Cascio, W. F. (2002). Responsible restructuring: Creative and profitable alternatives to layoffs. (San Francisco, Barrett-Koehler).

[45] Collins, J. (2001). Good to great: why some companies make the leap and others don't (New York: Harper Business).

[46] Thurber, J. (circa 1973). Unpublished study. Organizational Research and Development Department, Corning Glass Works, Corning, NY.

[47] Miller and Miller (2006) Managing for the Long Run, Boston, Harvard Business School 2006, (39) 13–17.

[48] Pfeffer, J. (1998). The human equation: building profits by putting people first (Boston: Harvard Business School Press) p. 67.

[49] Pfeffer, J. (1998). The human equation: building profits by putting people first (Boston: Harvard Business School Press) p. 67.

[50] Pfeffer, J. (2018). Dying for a paycheck: how modern management harms employee health and company performance (New York: Harper Collins).

[51] Brochet, F., Loumioti, M., and Serafeim, G. (2015). Speaking of the short-term: disclosure horizon and managerial myopia. Review of Accounting Studies 20 (3), 1122–1163.

[52] Chatman, J. A. and Cha, S. E. (2003). Leading by leveraging culture. California Management Review 45 (4), 29–31.

[53] Chatman, J. A. and Cha, S. E. (2003). Leading by leveraging culture. California Management Review 45 (4), 29–31, 72–73.

[54] Chatman, J. A. (1991). Matching people and organizations: selection and socialization in public accounting. Administrative Science Quarterly 36 (3), 459–484.

[55] Beer, M. and Weber, J. (2008). Asda (A1). Harvard Business School Case No. 498-006 (Boston: Harvard Business School Publishing).

[56] Jarnagin, C. and Slocum, J. W., Jr. (2007). Creating corporate cultures through mythopoetic leadership. Organizational Dynamics 36 (3), 288–302.

[57] Jarnagin, C. and Slocum, J. W., Jr. (2007). Creating corporate cultures through mythopoetic leadership. Organizational Dynamics 36 (3), 288–302.

[58] Beer, M. and Eisenstat, R. (2012). Becton Dickinson: opportunities and challenges on the road to the envisioned future. Harvard Business School Case No. 412-408 (Boston: Harvard Business School Publishing).

[59] Beer, M. and Eisenstat, R. (2000). Silent killers of strategy implementation and learning. Sloan Management Review 41 (4), 29–40.

[60] Groysberg, B. and Snook, S. (2005). Leadership development at Goldman Sachs. Harvard Business School Case No. 406-002 (Boston: Harvard Business School Publishing).

[61] Beer, M. and Weber, J. (2008). Asda (A1). Harvard Business School Case No. 498-006 (Boston: Harvard Business School Publishing).

[62] McCall, M. W. (1998). High flyers: developing the next generation of leaders (Boston: Harvard Business School Press).

[63] Beer, M., Finnstrom, M., and Schrader, D. (2016). Why leadership training fails–and what to do about it. Harvard Business Review 94 (10) (October 2016), 50–57.

[64] Amabile, T. (1988). A model of creativity in innovations in organizations. In Staw, B. M., and Cummings, L. L. (eds.), Research in organizational behavior 10 (Greenwich, CT: JAI).

[65] Beer, M. and Katz, N. (2003). Do incentives work? The perceptions of a worldwide sample of senior executives. Human Resource Planning 26 (3), 30–44.

[66] McCall, M. W. (1998). High flyers: developing the next generation of leaders (Boston: Harvard Business School Press).

[67] Binder, A. S. (ed.) (1990). Paying for productivity: a look at the evidence (Washington, DC: Brookings Institution).

[68] Beer, M. and Katz, N. (2003). Do incentives work? The perceptions of a worldwide sample of senior executives. Human Resource Planning 26 (3), 30–44.

[69] Beer, M. and Katz, N. (2003). Do incentives work? The perceptions of a worldwide sample of senior executives. Human Resource Planning 26 (3), 30–44.

[70] Beer, M. and Cannon, M. (2004). Promise and peril in implementing pay-for-performance. Human Resource Journal 43, (1) 3–20.

[71] Siegel, P. and Hambrick, D. C. (2005). Pay disparities within top management groups: evidence of harmful effects on performance of high technology firms. Organizational Science 16 (3), 259–274.

[72] Barron, J. N. and Kreps, D. M. (1999). Strategic human resources: frameworks for general managers (New York: Wiley). Chapter 3.

[73] Beer, M. and Katz, N. (2003). Do incentives work? The perceptions of a worldwide sample of senior executives. Human Resource Planning 26 (3), 30–44.

[74] Beer, M. and Cannon, M. (2004). Promise and peril in implementing pay-for-performance. Human Resource Journal 43, (1) 3–20.

[75] Barron, J. (2004). Commentary on Beer, M., and Cannon, M. (2004), Promise and peril in implementing pay-for-performance. Human Resource Journal 43, (1) 45–48.

[76] Barron, J. N. and Kreps, D. M. (1999). Strategic human resources: frameworks for general managers (New York: Wiley). Chapter 3, 96–97.

[77] Beer, M. and Spector, B. (1981). The Sedalia engine plant. Harvard Business School Case No. 481-148 (Boston: Harvard Business School Publishing).

[78] Lawler, E. E., and Hackman, J. R. (1969). The impact of participation in the development of pay incentive plans. Journal of Applied Psychology 53, 467–471.

[79] Brockner, J. (2017). The process matters. Princeton, NJ, Princeton University Press.

[80] Morgenstern, G. (2006). The best and the worst of executive pay. New York Times (September 17), https://www.nytimes.com/2006/09/17/ business/yourmoney/the-best-and-the-worst-in-executive-pay.html?mtr ref=www.google.com&gwh=81634DACB40CDFCAC1DE31F0F32BC FA6&gwt=pay

[81] Chatman, J.A. & Cha, N.E. (2003) Leading by Leveraging Culture. California Management Review, 45 (4) 25–42.

[82] Chatman, J.A. & Cha, N.E. (2003) Leading by Leveraging Culture. California Management Review, 45 (4) 31.

[83] Chatman, J.A. & Cha, N.E. (2003) Leading by Leveraging Culture. California Management Review, 45 (4) 31–32.

[84] Chatman, J.A. & Cha, N.E. (2003) Leading by Leveraging Culture. California Management Review, 45 (4) 31.

[85] Cascio, W. F. (2002) Responsible restructuring: Creative and profitable alternatives to layoffs. (San Francisco, Barrett-Koehler), 49–79.

[86] Schulze-Cleven, T. (2017). German labor relations in international perspective: a model reconsidered. IAQ-Forschung. http://www.iaq.uni-due.de/iaq-forschung/ The report number is 2017-07.

[87] Steak n's Shake (B) Case N1 405-0345 Harvard Business School Case, Boston, MA.

[88] Sasser, E., Schlesinger, L., and Heskett, J. (1997). *The Service Profit Chain*, The Free Press.

[89] Ton, Zeynep (2014) The good jobs strategy: How smart companies invest in employees to lower cost & boost profits, Seattle, WA, Lake Union Publishing.

# 2

# Managing Green Recruitment for Attracting Pro-environmental Job Seekers: Toward a Conceptual Model of "Handicap" Principle

### Do Dieu Thu Pham[*] and Pascal Paillé

Department of Management, Faculty of Business Administration,
Université Laval, Canada
E-mail: do-dieu-thu.pham.1@ulaval.ca; ppaille72@gmail.com
[*]Corresponding Author

Using signalling theory, this chapter aims to explore how signalling corporate environmental performance (CEP) during green recruitment has a positive effect on job seekers' perceptions of organisational attractiveness for environment. By reviewing literature on green human resource management practices and empirical studies, this chapter addresses the role of employee participation in environmental performance, displaying various kinds of pro-environmental behaviours. Accordingly, green recruitment is vital to securing and sustaining both quantity and quality in the workforce. This concept refers to a process of headhunting, stimulating and selecting qualified candidates who are sensitive to environmental sustainability and are willing to commit to environmental performance. Adapting the signal-based mechanisms by Jones, Willness and Madey (2014) and the CEP construct by Trumpp, Endrikat, Zopf and Guenther (2015), we propose a conceptual model which demonstrates that signals about CEP – in the form of environmental management performance and environmental operation performance – are positively linked to job seekers' perceived organisational attractiveness for environment via perceived organisational prestige/anticipated pride, perceived value fit and perceived favourable treatment. Also, we propose two new concepts – perceived signal honesty and perceived signal consistency – which reinforce the attracting mechanisms. The value of this model is twofold. First, it displays the validity of CEP that takes into account employees and the implementation level of environmental goals. Second, honest signalling about CEP during recruitment is worth practising for candidate attraction, which leads to multiple recruitment outcomes, such as job pursuit intention, job acceptance and job recommendations. Theoretical contributions and practical implications will be presented.

## 2.1 Introduction

Sustainability has long been defined as simultaneously pursuing economic, social and environmental goals – the so-called "triple-bottom line" (TBL) in which the environmental bottom line enables the regenerative capability of an ecosystem [1, 2]. However, TBL has been argued to be inadequate for organisational contributions to sustaining the Earth's ecology [3]. The conceptualisation of the TBL is rather rhetorical, as the equal achievement of the three goals is impossible, not to mention that it is unable to measure this TBL construct [3, 4]. As such, this paper emphasises the environmental sustainability (ES) goals and practices not exclusively to enable organisations to contribute to the regeneration of the ecosystem and natural protection, but also to improve organisational outcomes.

Scholars have pointed out that pursuing ES does not affect a company's economic development, but rather brings benefits that improve both economic and non-financial performance, including organisational performance, organisational reputation and competitive advantages [5–7]. On the one hand, the implementation of environmental management systems (EMS) helps reduce operational costs and charges in ways that ensure efficient consumption of energy, materials and resources, minimising negative environmental impacts and leading to positive financial outcomes [8]. Relatedly, green innovation of products and services allows organisations to create eco-friendly image and to gain agreement from customers. On the other hand, corporate environmental sustainability in management and operation contributes to organisational attractiveness, as it fulfils instrumental, socio-emotional and psychological needs of employees, especially those who care for natural protection and desire to have a significant impact through their work [7, 9, 10].

In the context of corporate greening, strategic human resource management (SHRM) is aimed to improve organisational performance consistent with environmental goals. Hence, the greening of SHRM results in green human resource management (GHRM), responding to the needs of an organisation regarding socio-environmental aspects. This intersection has been studied for years by scholars of human resource management (HRM); however, the concept of GHRM emerged and became broadly studied at a more proactive level in 2008. GHRM involves both traditional human resource (HR) practices and the more intangible ones, such as organisational culture, organisational learning, teamwork and employee empowerment, in alignment with environmental goals [11–13]. GHRM is designed to improve organisational resources and capabilities vis-à-vis environmental sustainability as well as to establish shared ecological organisational value and shared eco-knowledge among members of the organisation. Successful implementation of GHRM requires the involvement of entire staffs, rather than exclusively managers and specialists [13]. As such, green HR practices are designed with the purpose of motivating employees to participate in environmental operation and environmental management (EM) through which they introduce ecological concerns and make suggestions for environmental improvement, facilitating the spread of ethical values throughout the organisation [8, 13, 14]. This leads to the necessity of recruiting eco-minded employees who will contribute to the achievement of ES goals. Among the antecedents of organisational attractiveness for environment, CEP is an indicator of environmentally responsible employer. The purpose of this chapter is to explore how signalling CEP has a positive effect on job seekers' perceptions of organisational attractiveness. The objective of this chapter is threefold. First, the literature and recent empirical studies on GHRM will be reviewed to highlight the importance of employee involvement in the implementation of organisations' environmental performance. Second, this chapter uses signalling theory to explore how to form job seekers' perceptions of organisational attractiveness in green recruitment process. Third, a conceptual model is to be built to contribute new insights into the attracting mechanisms.

This chapter is organised as follows. We first review the literature on employee participation in environmental performance to address the importance of pro-environmental workforce, through which individual eco-mindedness and environmental identity will be highlighted. Green recruitment practices are then discussed through the lens of signalling theory with an emphasis on attracting pro-environmental job seekers at the very beginning episodes. Next, the effect of signalling CEP before and during green recruitment will be analysed by signal-based mechanisms. A conceptual model associated with our 10 propositions will both give an insight into attracting mechanisms and shed light on the "Handicap" principle for attracting pro-environmental ones. Finally, theoretical contributions, practical implications and suggestions for future research will be presented in our conclusion.

## 2.2 Theoretical Background

### 2.2.1 The Role of Pro-environmental Employees and their Participation in Environmental Performance

Pro-environmental employees (also called eco-friendly employees or green employees), in the form of fully skilled employees and talented managers, could be a source of the firm's sustainable competitive advantage when they are valuable, rare, imperfectly imitable and non-substitutable [5]. The resource-based theory views each individual as a link to a number of resources, including capabilities, explicit and tacit knowledge, good values and social capital, which can be a valued intangible asset that the organisation uses to implement their strategies [15]. From this view, the pro-environmental workforce and their participation in the firm's environmental performance is deemed crucial to the implementation of environmental goals. A pro-environmental workforce should be therefore understood to comprise talented executives and managers, skilled employees and workers who have ecological values, environmental identity, and are willing to commit to environmental performance. However, pro-environmental employees and their resources are valuable only if they are deployable in the workplace. Accordingly, their positive instrumental and psychological participations determine the implementation of environmental performance. Meanwhile, their motivation for environment can trigger their pro-environmental behaviours at work, which is considered a value-added to the organisation.

#### 2.2.1.1 Management role in environmental performance

Depending on the firm size and organisational structure, each company has different managing positions, differentiating among the leader, executive, middle-manager, supervisor or senior employee. These ones seem to resemble each other in their nature of management role. In this section, those positions are syncretised to be manager. Practically speaking, managers are the gatekeepers to environmental performance [16] because they are the ones who translate the environmental strategies into smaller goals and action plans, assigning green tasks to their followers, monitoring, supporting and motivating employee participation. More importantly, a pro-environmental manager is a good mirror for the followers. First, the manager is often seen as the representative of the organisation and thus, this one is a source of information that links to the appropriate behaviours [17]. Consequently, managers' pro-environmental behaviours, the in-role or discretionary ones, can directly or indirectly influence the pro-environmental behaviours of the subordinates. Second, manager's pro-environmental behaviours carry their values [13, 17] and in some cases, their personal values, rather than a widely applied rule, determine their green leadership behaviour and decision-making process [13]. In this regard, one quantitative study by Robertson and Barling [18] found that leaders' green descriptive norms predicted their green transformational leadership and pro-environmental behaviours, which later predicted employees' harmonious passion for environment. Interestingly, this finding led to the authors' assumption that what the leaders' friend and family do could spill over to the organisational context. Third, besides their model role, the pro-environmental managers act as the motivator for employees' pro-environmental behaviours [16, 18]. When employees feel inspired, psychologically empowered and experience a positive emotion derived from supervisor's/organisation's support, they will exert themselves to try to perform the expected workplace pro-environmental behaviours. In sum, those findings underlie the importance of the transformational style of the managers in shaping and driving an employee's workplace pro-environmental behaviours. The main purpose is to develop specific organisational pro-environmental behaviours and ecological values that carry the norms, beliefs and attitudes toward the environment.

### 2.2.1.2 Employee participation in environmental performance

While environmental strategies aim at environmental performance, GHRM practices aim at the participation of pro-environmental employees. Employee participation is demonstrated by the exhibition of pro-environmental behaviours (the so-called green behaviours), of which individual pro-environmental attitude, norm and perceived behavioural control (the so-called organisational affects) are important predictors [19].

Since a pro-environmental employee is also a private citizen, his/her attitudes and values influence work attitude and behaviour and vice versa [20]. Employee's pro-environmental attitudes reflect his/her belief toward ES issues. As such, the attitudes refer to the employee's overall assessment of the advantages and disadvantages of performing a given behaviour [19], for example, an eco-minded employee often weighs the consequences of a given behaviour for preventing its negative impact on the environment. As proposed by the theory of reasoned action and theory of planned behaviour (TPB) [21], individual attitude, subjective norms that derive from social norm/normative beliefs and perceived behavioural control are the three components in the prediction for individual behavioural intention. However, these theories have their limit in explaining relationship between behavioural intention and individual's real behaviour. Moreover, a gap exists in the prediction for the actual behaviour by personal norm. Personal norm, though also derived from and influenced by social norm, reflects one's self-concept, meaning the expectation that people hold for themselves. In this regard, personal norm embeds in it the individual's interests, characteristics and history. There are several contemporary scholars who are in support of this point of view and developed the predictors that extend the TPB model.

Lülfs and Hahn [19] proposed a model for predicting employee's voluntary pro-environmental behaviour of which the three predictors are attitude, perceived behavioural control and personal moral norm, which derive from individual awareness and social norm. Importantly, the authors assumed such behaviour to be determined by personal predispositions; hence, it is rational and purposive depending on: (1) employee's awareness of the need to proceed the action for problem-solving in a given situation, (2) his/her awareness of the consequences and (3) the extent to which he/she believes in his/her ability to engage in the action. Likewise, their model proposes individual habit as moderator between intention and behaviour, the authors assumed that habits at work would be able to change because they are generated by successfully performing stable behaviours in stable situations. Eventually, self-efficacy and self-esteem form an employee's attitude toward his/her engagement in a pro-environmental behaviour. It can be concluded that when personal tacit knowledge and capabilities toward an environmental issue are strong, the effect of employee's awareness is higher than that of social norm in predicting for the personal norm.

Similarly, the results of a quantitative study by Chou [22] implicated that personal environmental norm is more powerful than individual environmental beliefs in predicting for the employee's pro-environmental behaviour. Additionally, employees with more positive pro-environmental attitude and higher personal environmental norm tend to consistently engage in pro-environmental behaviours, regardless of weak green organisational climate [22] or their level of experienced positive daily affect, which is the individual's emotional experiences about any target they may encounter in the day [23]. Another quantitative study by Dumont et al. [24] also indicated that individual values moderate the effect of psychological green climate on extra-role green behaviour. These research findings lead to the conclusion that there should be a compatibility between individual values and organisational values as well as the compatibility between personal environmental norm and perceived environmental behavioural control for the highest positive workplace outcomes.

There are a number of ways in which employees can engage in acting for ES at work, exhibiting various kinds of pro-environmental behaviours. Accordingly, firms must be able to rely on their employees for both hands-on practices [25] and hands-off practices such as environmental problem prevention, energy and material saving, waste reduction, cleaner production, green workplace, etc. Moreover, pro-environmental employees differentiate from low-intensive to high-intensive depending on the level of the individual's environmental values and environmental identity. In the organisational setting, Ciocirlan [20] has refined such kind of behaviours, namely environmental workplace behaviour, to be composed of organisational citizenship behaviour for environment (OCBE), environmental in-role behaviours and environmental counterproductive workplace behaviour (ECWB). This paper exceeds the previous research in the degree to which it accounts for the ECWBs and the situations when the environmental values of employee are stronger than that of the organisation. ECWB is assumed to not always be unexpected or unethical, as environmental workplace behaviours may overlap with ECWBs including, for example, the silent behaviour, disobeying behaviour to order or speak out behaviour of employees to oppose certain rule/policy/behaviour that they perceive as harmful to the environment. Besides, ECWB is, at times, associated with higher risk-taking behaviour, higher creativity, higher self-confidence, higher liberty, less normative behaviour or power-influencing behaviour.

OCBE, often known as the extra-role behaviour, is the most recently researched in field. What makes OCBE different from the voluntary pro-environmental behaviour, which is studied by Lülfs and Hahn [19], is that it cannot be coerced by contractual terms or punishment, it is more discretionary and based on rational choice. An example of this kind of behaviour is helping behaviour so that it is associated with the prosocial ones which are for the welfare of others. In the one part, individual positive environmental values and OCBE will be generalised within the workplace and construct the collective OCBE, which refers to "the perception of what is considered the standard mode of behaviour in the unit with regard to environmental matters" (p. 202) [26]. In the other part, OCBEs facilitate the effectiveness of EM and environmental improvement and thus, are highly appreciated by organisations. First, the participation of employees in environmental initiatives is theoretically and empirically proved to be positively associated with higher engagement with the organisation, organisational performance and job satisfaction, which are negatively related to intention to quit [27, 28]. Second, socialisation, altruism and helping among organisational members increase firm performance [29].

In this respect, the social network that enhances individual performance is not exclusively built by a single, dyadic relationship, or leader–member exchange relationship, but also on a broader framework, where the person receives both internal and external supports. The study by Bruque et al. [29] surveyed 371 employees working at 133 different branches of a large financial firm in Spain in the context that the firm made a major change to its Information System. Results indicated that supportive social ties had a positive effect on individual's organisational citizenship behaviour, which was directly associated with individual task performance and adaptation to change. As such, it can be concluded that OCBEs and individual environmental identities, which reflect their social network, are crucial to the successful implementation of environmental activities and thus highly appreciated by an organisation, especially when it goes through a difficult time or an organisational change.

### 2.2.2 Green Recruitment through the Lens of Signalling Theory

### 2.2.2.1 Overview on organisational attraction in green recruitment

Given the importance of the enduring participation of pro-environmental employees in environmental performance and the spread of ecological values within the organisation, green recruitment plays a foremost role in attracting and recruiting pro-environmental job seekers. Green recruitment is referred to as a process of headhunting, stimulating, recruiting and selecting candidates who are sensitive to ES and willing to commit to environmental issues [8, 30]. A body of works that have made a great contribution on this field includes Gully et al. [7], Jepsen and Grob [31], Milliman [32], Uggerslev et al. [33], etc. The recruiting practices involve green employer branding, which refers to organisational prestige related to EM and can be formed through GHRM practices, while the selecting process deals with a series of tests, interviews and candidate evaluations based on green criteria [8, 30]. It is therefore crucial that organisational attraction practices are embedded in green recruitment process to attract and sustain the interest of pro-environmental candidates toward the organisation and the job vacancy, especially in the "war for talent". In assuming that job seekers' first impression on employer influences their interest in later stages of green recruitment, in this chapter, we highlight the organisational attraction practices at the initial stages for attracting and generating prospective pro-environmental applicants.

Organisational attraction is referred to as affective and attitudinal positive thoughts of job seekers or participants about an organisation as potential employers [7]. Hence, organisational attraction practices concern the organisation's capacity to attract its job seekers [34] and then to become an employer of choice to prospective applicants. For the highest effectiveness, an organisation needs prerequisite capabilities to provide prospective job seekers good offers. A study by Renaud et al. [34] using a "policy-capturing" approach investigated on the effect of symbolic organisational attributes and instrumental organisational attributes on applicant attraction and found that the symbolic had the strongest effect. Such that, the participants who rated in a scenario of "ethic is important for the organisation" reported a higher level of attraction than those who rated in the scenario of "ethic is not important for the organisation". The findings confirmed the combined effect of instrumental and symbolic organisational attributes on applicant attraction, suggesting a firm to create a total reward program and code of ethics. Given that environmentalism is a sub-set of business ethics, implementing ES and having ecological values are worthy to attract eco-minded job seekers. Implementing ES practices signals that the company is not only responsible for its operation but also for the society and its stakeholders, making the organisation attractive to customers and potential employees. Among the corporate sustainability practices adopted by companies, CEP and its sisters – corporate social performance (CSP) and corporate social responsibility are frequently studied in relation with organisational attraction, and are found to be indicators of organisational attractiveness [6, 35–38]. The intersection of the three constructs addresses the environmental issues, representing the harmony between economic, social and environmental sustainability goals of development. CEP differentiates from CSP and corporate social responsibility in the extent that it necessarily aims at ES goal. CSP encompasses a firm's self-regulation codes, standards, ethics and norms that are integrated into the business and organisational model [37], so does CEP.

CEP has long been studied and often used interchangeably with corporate environmental sustainability or corporate environmental responsibility; however, we contend that Trumpp et al. [39] provided the most complete conceptualisation, which relies on ISO definitions. CEP is assumed to focus on both EM activities and the outcomes of these activities and processes. Accordingly, Trumpp et al. [39] defined CEP as a multidimensional aggregate construct, which is composed of at least two dimensions, namely environmental management performance (EMP) and environmental operation performance (EOP). While EMP encompasses the EM activities and refers to the strategic level of environmental performance, EOP refers to the operational level and therefore, can difficultly be measured in all-inclusive manner.

This new conceptualisation highlights the involvement of employees at all level in environmental performance. Relatedly, pro-environmental employees constitute a great proportion of the success of environmental performance. Recruitment is a process whereby the recruiters and the applicants attempt to detect and evaluate each party's values/characteristics based on the information they perceive before and during the recruitment stages. It is apparent that within the very beginning episodes of green recruitment of which the main purpose is to attract pro-environmental job seekers to generate a larger pool of pro-environmental applicants, the employer will strive to send the information on what is expected by the pro-environmental job seekers or on what they are looking for. In this chapter, we define a pro-environmental job seeker to be a person who is sensitive to ES and is currently seeking for a job or has the tendency to go to the job market. Inevitably, sending the information on CEP while recruiting has a potential to interest these pro-environmental job seekers.

### 2.2.2.2 Signalling effect in green recruitment
#### 2.2.2.2.1 *Overview on signalling theory*
Signalling theory, a multidimensional approach to the nature of human and nonhuman signals, is applicable to a variety of fields of study, including management and HRM [40]. Signal is defined as a sign or an indication of something that conveys notice or warning or transmits information (Merriam-Webster dictionary); it therefore can be recognised by a multitude of senses: seeing, hearing, smelling, touching and tasting that constitute a thought/feeling and generate one's perception. Signals are to reveal the nature, underlying quality or goals of a human, nonhuman or a phenomenon (e.g., a report on ES activities indicates a company's rate of greening orientation and performance, company's revenue or investment signals its financial standing, a letter of recommendation by a prestigious person indicates the reliability and appreciated competence of the referenced applicant).

Signals often carry a message from a signaller to a receiver with the intention of forming the receiver's feeling or belief on the signaller's characteristics or goals [40]. To assure the purposeful message be successfully interpreted by the receiver, the signal should be perceived as honest. Honest signalling and its effective outcome are vitally important in building and sustaining a perception and relationship between signaller and receiver. It is crucial that the signaller has the underlying quality that is associated with the signals sent and that the signal should be costly or hard-to-fake for it to be perceived as honest [41]. In this regard, the "Handicap" principle is a metaphor of the honest signalling, the key concept in signalling theory.

"Handicap" principle implies that sending a message on accurate information imposes a cost on the sender that only certain individuals can bear [41]. The high cost the signaller takes on can be their money, time, energy, effort or sacrifice. Otherwise, the "Handicap" stands for an explicit incapability or a loss which is often perceived as a true signal because of less probability of being manipulated. This is the reason why a costly signal is termed interchangeably with signal fit, signal reliability, signal veracity and hard-to-fake signals. A signal is reliable because only the truly high-quality signallers can bear the cost. For example, a thorough hand-written application letter is an honest signal, as it demonstrates the candidate's motivation toward the job and the employer, as well as the candidate's past experiences, which are hard to fake. It also shows that the candidate took the time to write the letter in a personalised style, which is not copy-and-paste elsewhere. From the recruiter's perspective, a letter of invitation and feedback composed to target a specific individual is always eye-catching and preferable to a letter format sent to everyone. A direct phone call to the applicant may infer his/her perception that he/she is admired and welcome. This is also effective in cases when an applicant is rejected, as it demonstrates the employer's gratitude for the time and interest that the participant dedicates to the job vacancy and therefore produces a perception of thanks. However, signals could be in the form of fixed attributes or the manipulable attributes, which are in a shifting conditional probability distribution to form the employer's beliefs. As such, effective signalling is based on an adequate number of signals within the appropriate cost range [42]. Moreover, a hard-to-fake signal is more likely perceived as honest than the costly signal, because the former is beyond the conscious control of the individual and is, therefore, intrinsically difficult to manipulate [41]. For example, graphology, voice or speech may be applied as kind of personality test for candidate selection because it is hard to fake. Similarly, licence and patent acquisition is an honest signal representing the acquirer's underlying quality because it is hard to fake. Importantly, the "Handicap" is condition-dependent since only high-quality signallers are capable of meeting the requirement.

Signalling effectiveness can be enhanced by signal frequency and signal consistency [40]. Signal frequency refers to the extent to which more observable signals are produced or to which signals are increased in number. Due to the dynamic environment of signalling systems, the information available is constantly changing; signal consistency by which one sends various signals carrying the same message can therefore strengthen the effectiveness. In addition, signalling effectiveness is determined in part by the characteristics of the receiver. It is crucial that the receiver heeds the signals and interprets them in the manner that fits the signaller's intent. Since the message is subject to be coded, different receivers may translate the signal to have different meanings, indicating that receivers' interpretations are important in signalling effectiveness. In fact, signalling is a two-way information exchange between the sender and the receiver. The receiver should, therefore, send feedback to signallers about the effectiveness of their signals. Sometimes, receivers can give feedback in the form of countersignals or camouflage to facilitate efficient communication [40]. Relatedly, attention to countersignals can result in more efficient signalling in the future. In such a case, both the signaller and the receiver desire to obtain information about each other, for attaining the effectiveness, they may learn to improve the reliability of signalling by studying which signals are more reliable, which ones get more notice from the receiver and how the receiver interprets these signals. To maintain the effectiveness of signalling, the costs of signals must be structured in the way that false and misleading signals cannot pay; otherwise, they would be ignored by the receiver [40] or, in some cases, the receiver may give negative feedback or punishment, called a costly penalty.

The nature of signalling theory is that the signallers are the insiders (e.g., CEO, managers, recruiters) who know the information that is not available to the outsiders (e.g., stakeholders, job seekers) who want to explore and obtain that information. Additionally, there are often multiple signallers or signals so that information asymmetry exists as an inevitable phenomenon. As a result, signalling should aim at obtaining an unobservable quality and reducing information asymmetry. Furthermore, various signallers may benefit differently from the receiver's response, as they will choose different sizes of the "Handicap" to produce. Each one tends to signal in the "Handicap" manner most appropriate to themselves to make it easier for the receiver to judge the sender's quality. As a result, the "Handicap" principal is becoming more complex [41].

### 2.2.2.2.2 *Signalling effect in attracting pro-environmental job seekers*

Generally speaking, recruitment is a mind-reading process between the job seeker and the recruiter because each party lacks information on the other; hence, they will assess the unobservable based on the observable [41]. They will also use what information they have, for example, gender, income, experience, university transcripts and so on, to infer the characteristics of the other [42]. From the applicant's perspective, signalling theory explains that job seekers have incomplete information about the employer [40, 43], and the information supplied by recruiters during selection process serves as signals about organisational attributes and job attributes to inform applicants about their would-be life at that organisation if they are hired. Conversely, recruiters conclude applicants' capabilities and fit with the job by studying their résumés, assessing job interview responses and conducting selection tests on skills and personalities. This signalling is therefore an interaction between job seekers and employers or recruiters that may occur at job fair, job interview or company branding activities, what is called by Bangerter et al. [41], a signalling game that depends on the job market pressure. In the context that there are more and more companies engaging in ES and that the job seekers might have a perception of *greenwashing[1]* toward the organisation, the pressure on the employer in the signalling game is higher than on the pro-environmental job seekers, especially at the beginning episodes of green recruitment. Moreover, in the "war for talent", the employers tend to take part in the "arms race" to compete with each other to win the employment with the prospective employees. As such, the differentiation and the reliability of the pro-environmental employer/recruiters or the information source are crucial in attracting pro-environmental job seekers. This chapter assumes that "Handicap" principle characterising the costly or hard-to-fake signals and its derived characters is the principle of green recruitment.

To date, because of the vast development of information technology and the Internet, people worldwide go online more frequently to communicate and search for information. Pro-environmental recruiting message on internet associated with e-recruitment is therefore an effective signalling channel that quickly catches the attention from pro-environmental job seekers. In this regard, Kashi and Zheng [44] assumed that the job seekers perceive anything or anybody related to the organisation as signals of how they feel as being working in that organisation. The impression on company's recruiting web page therefore forms their impression on the job vacancy and on the employer. Accordingly, people are often attracted to what they concern about, the pro-environmental job seekers, who are sensitive to ES, will be attracted to such pro-environmental information. Examples can be

---

[1]*Greenwashing is when a company promotes environmental performance and images, but operates in an opposite manner to the goal of the announcement, or when it utilizes exaggerated advertisement in an unbelievable way (http://www.investopedia.com/terms/ g/greenwashing.asp).*

drawn from the quantitative study by Behrend et al. [45], which highlighted the effects of pro-environmental recruiting message in attracting prospective employees. In believing that applicants who are highly concerned for the environment are more likely to pursue employment and accept job offers from pro-environmental organisations, the authors hypothesised that the pro-environmental message (e.g., on the company website) would affect the job pursuit intention of job seekers who were relatively supportive of the environment more than it would affect the preferences of those who were relatively unsupportive of the environment. Interestingly, the results rejected this hypothesis; the message's effect on job pursuit intention was not dependent on individual's personal environmental stance. Results showed that the effect of the pro-environmental message was significant, such that participants who saw this pro-environmental job advertisement were willing to pursue employment with the signalling company. Remarkably, the mediating role of organisational green reputation in the relationship between the pro-environmental message and pursued employment was supported. Behrend et al. [45] explained that those individuals interpreted the organisation's concern for the environment as a sign that it is caring, trustworthy and would therefore show concern for its employees. Additionally, these job seekers thought that if the organisation could spend money on environmental activities, it could afford to pay its employees well.

Another effective signalling channel is the recruiter/HR specialists as these ones are considered the employer's representative and a source of information that links to the organisation. In this respect, scholars on applicant attraction have linked recruitment with impression management (IM). One qualitative study by Wilhelmy et al. [46] explained the signalling process in an interview from the viewpoint of the interviewers in order to find how and why interviewers tried to make impressions on applicants. The authors assumed that the interviewer strived to detect what interests the applicants to send appropriate signals to deliberately create impressions on them. Their results revealed two types of IM intentions, the primary refers to the goals of representing the organisation by signalling organisational attractiveness and authenticity, while the secondary refers to the interviewer's personal interaction with applicants, including signalling closeness (with the purpose of building rapport, appreciation or trustworthiness) and signalling distance (to demonstrate the professionalism or superiority). Both closeness and distancing can be adapted simultaneously by the interviewer. To exhibit those intentions, the interviewer can adopt from numerous behaviours categorised into three groups: (1) verbal IM behaviours, whereby interviewers focus on the content to influence applicant impression; (2) paraverbal IM behaviours, which are verbal behaviours other than words, such as modulating the voice; (3) nonverbal IM behaviours, which refer to body language; (4) artifactual IM behaviours, which concern how interviewers use an object/other aspects to influence the impression, such as their appearance, visual display during interview, giveaways or promotional items for applicants or seating arrangements and (5) administrative IM behaviours, which refer to the timing communication and services, such as offering drinks, travel expense refunds, personal invitations and receipt confirmation of application documents. The findings showed a variety of intended IM outcomes in recruitment, including valued information cultivated from applicants' personal disclosure, applicants' positive attitudes and perceptions toward the organisation, stronger reputation and career advancement of interviewers, applicants' job choice intention and behaviours and applicants' recommendation intentions and behaviours. In green recruitment, it is necessary that the recruiters/interviewers behave in a pro-environmental manner unified with the organisation's pro-environmental stance, because the fact that they "walk their talks" will leave the participants a good impression on the organisation's ecological characteristics or organisational green culture. Additionally, to increase the signalling effectiveness, another study by Wilhelmy et al. [47] have confirmed the positive effect of organisation-enhancement, the extent to which the recruiter strives to create an attractive image to "sell" the organisation to the applicant or to enhance the organisational prestige, and applicant-enhancement, the extent to which the recruiters show the interests for the applicant's ideas or admirations on his/her past achievement/capabilities. Notably, consistent with signalling theory perspective, Wilhelmy et al. [47] also found that the degree to which the signalling is successful depends upon whether the job seekers actually receive the signal and that this could be studied via applicant's reactions and perceptions.

As aforementioned, the reliability of the green signals determines the job seeker's positive perception toward the organisation. We assume that information from third party will increase the reliability since it reduces the subjective opinion. From the job seekers' part, their relatives, friends, colleagues, acquaintances and the associates are a good address to seek for further information and opinion. As such, the employer can make use of this and signalling on green recruitment via their employees. Taking this into consideration, the employer can provide employee testimonials (video or image) whereby the extant employees share their experiences and opinions toward the pro-environmental issues and their job. Otherwise, the employees/recruiters can play their part as the information forwarder. Their social network will be then a potential pool of candidate. This signalling manner is effective particularly in chasing the best fit because the employee/recruiter is the middleman who knows well both parties (the employer and the desired candidate). For both cases, we anticipate that selling environmentally-related personal story and experiences can attract the pro-environmental job seekers because these ones seek for the real context to make inferences on their would-be life at that organisation and have the tendency to believe in personal sharing.

Above all, a pro-environmental job seeker is both a potential employee that links to a pro-environmental network and a potential customer/client. Successful signalling in green recruitment will be associated with pro-environmental job seeker's good perception toward the employer, the recruiter or the job vacancy and therefore, brings multiple outcomes, including but not limited to job seeker's intention to apply, job pursuit, job acceptance, job recommendation or their intention to buy/recommend the company's products/services. One cannot disregard entirely the surroundings and often pursues the signals to which he/she is attracted. Therefore, pro- environmental signalling is useful and worthy for attracting pro-environmental job seekers, provided that the employer's/recruiters' behaviours comply with the "Handicap" principle.

## 2.3  Proposition Development

### 2.3.1  Linking CEP and Organisational Attractiveness for Environment through Signal-based Mechanisms: A Conceptual Model of "Handicap" Principle

One study by Backhaus et al. [36] studied the participants' ratings on the most popular dimensions of CSP and found that the environment was among the three most powerful factors to inform organisational attractiveness as an employer; the other two were community relation and diversity. This finding supports the idea that job seekers have long been sensitive to information on the environmental activities of the employer, which signal its prestige, ecological values and prosocial orientation. What needs to be further studied is how candidates interpret the various signals on CEP and how their perceptions are transferred to perceived organisational attractiveness for environment.

We propose discussing to what extent five signal-based mechanisms (including job seeker's perceived organisational prestige/anticipated pride,

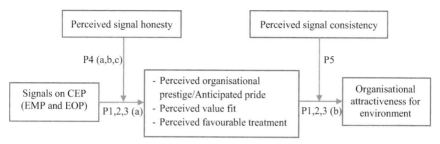

**Figure 2.1**   Proposed conceptual model inherited and developed from Jones et al. (2014) and Trumpp et al. (2015).

perceived value fit, perceived favourable treatment, perceived signal honesty and perceived signal consistency) intervene in and reinforce the relationship between CEP – in the form of EMP and EOP – and job seeker's perceived organisational attractiveness for environment. We assume that signalling theory and its "Handicap" principle are suitable to explain the process by which job seekers form their impressions on the employer, which may lead to other recruitment outcomes, such as job pursuit intention, job acceptance, job recommendation or the increase in organisational green prestige. Figure 2.1 is our proposed conceptual model inherited and developed from Jones et al. [48] and Trumpp et al. [39].

### 2.3.1.1 Perceived organisational prestige/anticipated pride

Grounded in social identity theory, job seekers' anticipated pride results in their identification with the employing organisation due to the feeling that becoming an employee of this organisation will enhance their self-worth [48]. Organisational prestige refers to the perception of the society and stakeholders on an organisation and thus corresponds to the organisational reputation [49–51]. That is, when a company is considered prestigious, it has social approval for its business operation and gains trust from its stakeholders. Participating in ES practices contributes to organisational reputation because it introduces the prosocial orientation of the organisation and responds to normative expectations of stakeholders [50]. For example, CEP pertaining to an EMS (e.g., ISO 14001) signals the firm's economic and legal responsibility.

In addition to an action plan, the legitimacy of an environmental policy and the strategic management of environmental performance gain the credibility of organisational prestige. Jones et al. [52] found results supporting their hypothesis that an employer's ES practices send signals to job seekers about the employer's reputation and prestige, from which job seekers infer how proud they would feel as one of its employees. As a result, signalling on such practices or on CEP, as formally assumed in this chapter, generates job seekers' anticipated pride and, in turn, their perception of organisational attractiveness for environment. Additionally, pertaining to the characteristics of signal frequency, the richer is the information on CEP, the more effective the signalling mechanism will be. As such, we propose as follows:

*Proposition 1a: Perceived signals about the employer's CEP, in the form of EMP and EOP, are positively linked to job seeker's perception of organisational prestige or anticipated pride with the organisation.*

*Proposition 1b: The job seeker's perception of organisational prestige or anticipated pride is positively linked to their perceptions of organisational attractiveness for environment.*

### 2.3.1.2 Perceived value fit

The conceptualisation of person–organisation fit (P–O fit) posits that a person is apt to perceive a fit with an employer if he/she finds the compatibility between himself/herself and the organisation with respect to values, goals or traits [53]. In this context, the P–O fit is subjective compatibility because it is based on the candidate's own evaluation; for example, *"My values correspond to those of the employees at this organisation"* or *"I think that my personality corresponds to that of the employer/organisational image"* [54]. Candidates tend to seek an employer who can fulfil their needs and meet their expectations on opportunities and personal attitudes on development. In this regard, P–O fit is believed to be a powerful predictor of organisational attractiveness.

An empirical study on 287 final-year undergraduate students at a large business school in a developed Asian economy, which emphasised on different types of expectations for P–O fit, proved the key role of individuals' expectations on both value expression and needs–supplies [55]. The author explained why job seekers care about a P–O fit on values and opportunities for value expression. Consequently, symbolic value, including organisational reputation, is one of the decisive factors of job seekers' acceptance behaviour. A meta-analytic test of seven commonly-explored predictors of applicant attraction found that P–O fit was the largest predictor, suggesting organisations find ways to foster candidates' perceptions of fit [33]. Also, Gully et al. [7] found that the positive effect between perceived P–O fit and organisational attractiveness was greater among individuals with high socio-environmental values. This finding suggests that employers should communicate their ecological values in their recruiting message. Also, the results of Jones et al. [52] supported another hypothesis that an employer's ES practices send signals to job seekers about its organisational values, from which they infer perceived value fit. Similarly, our propositions for the perceived value fit mechanism are as follows:

*Proposition 2a*: *Perceived signals about the employer's CEP, in the form of EMP and EOP, are positively linked to job seeker's perceived value fit with the organisation.*

*Proposition 2b*: *The job seeker's perception of value fit with the organisation is positively linked to their perceptions of organisational attractiveness for environment.*

### 2.3.1.3 Perceived favourable treatment

There are two reasons that lead to our assumption on the mechanism of job seeker's perceived favourable treatment. First, one of the two main goals of interviewers introduced in recruitment literature is applicants' positive emotions or making the interview a pleasant experience for applicants [47]. Accordingly, from an attraction perspective, people are often attracted to those who treat them good. In the context of green recruitment, we contend that both environmentally friendly settlement and ambience created by the employer's/recruiter's during the interaction with the pro-environmental job seekers as well as the employer's pro-environmental behaviour that are favourable to the pro-environmental job seekers make them have a positive emotion and notice the employer's green signals.

Second, based on the received signals, a job seeker might have an expectation on a favourable treatment which refers to his/her hope that the employer cares for its employees in a just manner and corresponds to his/her expectation of favourable employee treatment [38]. The positive relationship between ES practices and job seekers' expectation on favourable treatment was also supported by Jones et al. [52]. From a candidate's perspective, there is a dearth of information on working condition and an organisation's attitude toward its employees; candidates are often attracted to those signals which form their perception of how well the employer would treat them as its employee. CEP corresponds to four dimensions of CSP, including economic citizenship, legal citizenship, ethical citizenship and philanthropic citizenship, in which the economic dimension represents a firm as a basic economic unit in society, given that the firm cares for its employees [37]. This means that the employer is expected to provide a good working environment and a good employee life cycle, including recruitment, employee induction, appraisal and promotion [16]. There are some explanations for this mechanism. First, due to the legitimacy of an EMS, job seekers may interpret signals of organisational management and operation conforming to the environmental standards as a safe, eco-friendly working condition. Second, in terms of the philanthropic citizenship approach, when the employer is actively engaged in discretionary ES activities, this signals that it cares for the society and others' well-being. This generates job seekers' expectations of just treatment. Third, as earlier presented, CEP is positively related to financial performance because it not only reduces waste, energy and material consumptions, but also increases work efficiency, leading to higher organisational outcomes. When an organisation is engaged in ES practices, it means that it has enough resources to invest in and can afford these activities. As a result, job seekers may infer that the organisation has good financial standing and, in turn, can sustain those environmental practices. Working for such an employer will bring job seekers opportunities to fulfil their needs regarding pro-environmental issues. Taken together, we propose as follows:

*Proposition 3a*: *Perceived signals about the employer's CEP, in the form of EMP and EOP, are positively linked to job seeker's perceived favourable treatment.*

*Proposition 3b*: *The job seeker's perception of favourable treatment is positively linked to their perceptions of organisational attractiveness for environment.*

### 2.3.1.4 Perceived signal honesty

In this chapter, we assume that the employers'/recruiters' behaviours and the signalling game need to comply with the "Handicap" principle. Such that, the signalling and the signal itself should be costly or hard to fake and demonstrate the employer's underlying quality. On the one hand, this will increase the reliability and trustworthiness of pro-environmental signals. In this regard, a job seeker's positive point of view on certain signal must stem from his/her assessment on this signal/the signalling that is perceived as correct. Otherwise, the motive of the signal/signalling is perceived as for the right pro-environmental objectives, it can be also for the welfare of the employees, the community or for the job seeker's benefit. On the other hand, when a pro-environmental signal has a distinctive character or is differentiated from any others that are being produced by other employers, it will inherently eliminate the uncertainties and help the job seeker to identify easily the employer's pro-environmental characteristics. Additionally, from an attraction perspective, a strange signal has the likelihood to catch attention and at times, even a tiny one can have an attracting effect provided that it is outstanding. As a result, we assume that perceived signal honesty is vital in making inferences, especially at first sight because at the moment when the job seeker senses the signal, this one naturally has a first quick response – a thought even without conscious reasoning. Accordingly, the higher level is the job seeker's perceived signal honesty, the more effective is the signalling mechanism. Our propositions are as follows:

*Proposition 4a: The higher is the job seeker's perception of signal honesty, the more likely they perceive an organisational prestige for environment or anticipate a pride.*

*Proposition 4b: The higher is the job seeker's perception of signal honesty, the more likely they perceive a value fit with the organisation.*

*Proposition 4c: The higher is the job seeker's perception of signal honesty, the more likely they perceive a favourable treatment.*

### 2.3.1.5 Perceived signal consistency

Signal consistency refers to the extent to which all signals carry the same message/characteristic. As aforementioned, the signal consistency increases the signalling effectiveness. The nature of signalling game is that each party strives to get each other's information and make inferences about each one's characteristics. Apparently, a job seeker might wonder about the employer's characteristics. Naturally, the job seekers will chase the employer's history and make a link to what they have already known, which is referred to as internal search and memory scan, in order to make an overall assumption. A job seeker therefore goes through a retrospective process in which he/she may make evaluation or judgement on the employer's characteristics based on a combination of or a comparison between the newly received signals and those ones in the past events. In this case, the extent that all pro-environmental signals, which the job seeker receives before and during the green recruitment stages, are consistent will increase the likelihood that he/she has an overall positive perception toward the employer and perceive organisational attractiveness for environment. The consistency here can be referred to as the extent that the new behaviour is consistent with those behaviours that have been already observed [56]. Otherwise, we assume that a job seeker perceives a signal consistency when he/she finds that the pro-environmental signals are logic and that these ones are compatible with the employer's orientation, their ecological values or with their (new) pro-environmental objectives.

As previously analysed in the signalling game, it will be easier for the job seeker to judge the employer's quality if the employer signals in the "Handicap" manner most appropriate to themselves. As a result, we propose as follows:

*Proposition 5*: *The more the job seeker perceives the signal consistency, the more they perceive the organisational attractiveness for environment.*

### 2.3.2 Relevance of "Handicap" Principle in Explaining Attracting Mechanisms in the Context of Corporate Environmental Performance

Although previous literature has addressed the positive effect of CSP, including environmental dimension, on organisational reputation and organisational attractiveness [6, 35, 36, 48], little is known about the mechanism of how the implementation of CEP, at both the strategic level (EMP) and the operational level (EOP), has a positive effect on attracting eco-minded candidates. We contend that signalling theory and its principle contribute great value in recruiting new dedicated pro-environmental employees who have the consciousness of environment. Especially, the two new signal-based mechanisms proposed in this chapter along with other possible moderators discussed below will provide the insights into the attracting mechanisms that explain why a pro-environmental job seeker might or might not perceive a value fit with the organisation and how to activate his/her interests for the environment in general and for the organisation in particular.

Previous researches showed mixed support for the effect of the perceived value fit mechanism, especially in those studies using individual differences in pro-environmental attitudes. In the manipulating study of Jones et al. [48], senior undergraduate participants were asked to review fictitious company web pages, and the effects of perceived value fit were above and beyond those of the other two mechanisms. An analysis on individual differences also resulted in those with higher pro-environmental attitudes rating the organisation as being more attractive. Another supportive study found that perceived value fit positively related to both organisational attraction and job pursuit intention, and that the effect of communicating high levels of socially and environmentally-responsible goals were related to a high level of job pursuit intention among those with a stronger desire to have significant impact through their work [7]. On the contrary, other studies failed to support both the perceived value fit mechanism and the difference on personal attitudes toward the environment [6, 45].

There are some explanations for the unsupportive results on perceived value fit and the "Handicap" principle does assist. First, it may be due to the strong effect of a manipulated green recruiting message that leads participants to perceive organisational attractiveness in a different manner. For example, individuals who do not hold ecological values may still evaluate an environmentally friendly company positively (as shown in the results of the study by Behrend et al. [45]). In this case, the relationship between CEP and organisational attractiveness is not explained by the perceived value fit, but by other two signal-based mechanisms. Second, since the participants' perception of value fit is subjective – as they are focused on their comparison between their own values and the signals received from the recruiter or the researcher who manipulated the study – it depends on their regards toward the pro-environmental issues. Their attitudes or beliefs toward certain dimensions of CEP might be different from those assumed by the recruiter or researcher. Furthermore, individual ecological values differ across participants depending on two utility functions motivated by egoistic and altruistic considerations [10]. Egoistic individuals are more likely to engage in ES practices for their self-enhancement and personal benefits, while those with altruistic motivations are more for the welfare of others, specifically for the environment. Third, since companies differ in their ES practices, each participant may evaluate these companies differently. Not to mention, because those studies were all based on fictitious or manipulated recruiting messages (electronic or printout version), participants' interpretations and perceptions could vary depending on which information of the environmental construct had been used to signal in the survey and the size of the "Handicap" signals. For example, the supportive study by Jones et al. [48] signalled the effect of the environmental dimension of CSP using the information on donation and employee volunteer programs, employee involvement in energy reduction and recycling. These explicitly specific signals facilitate participants' interpretation. Also, the message could attract both egoistic and altruistic individuals. Contrarily, the study by Behrend et al. [45] tested the effect of a recycling symbol and a statement (company RLA supports the environment) without no other additive information on a company's environmental practices. This signalling limited participants' interpretation on ecological values. Additionally, the message highlighted the company's fast-paced environment, which is dynamic, innovative and fun. This accidently created information asymmetry between ecological issues and the physical working environment, while signalling theory aims to reduce information asymmetry. Finally, the unsupportive results on perceived value fit are those of pioneer studies using sophomore and college student samples [6, 45], while the supportive results came from recent studies using a university senior student sample or a large online sample of job seekers [7, 48]. Consequently, there may be a possibility that ecological values differ among different generations, levels of education and job experience. Likewise, as these studies are all conducted in the United States and Canada, the fact that the effect of the perceived value fit mechanism has become stronger in recent studies (conducted in 2012 and 2014) than in the previous (conducted in 2000 and 2009) may be due to a shift toward a pro-environmental stance and a more familiar approach that enables employee involvement and then enhances eco-mindedness.

Another unexpected result on organisational attractiveness pertains to the question of why some are attracted to information on a company's CEP practices and others are not. In the study by Jones et al. [52], a little more than one-third of the total sample claimed that information on community involvement and ES did not enhance organisational attraction. The responses to an open-ended question indicated that eight participants, accounting for 20.5% of subsample, perceived a general lack of fit; three participants prioritised other factors, such as compensation/pay and job role or promotion; two participants worried that investment in ES would detract from the company's profits and another participant thought it would have been better, if there had been pictures with the company's employees so that he/she could picture himself/herself working there. The authors [52] found factors leading to the scepticism and cynicism of job seekers, including prior experience with an employer's *greenwashing*, the need to see/experience the company's ES practices to believe it, the need for more detailed information and the questioning of the nature/motives they attributed to the employer's investment in ES practices. Briefly, most scepticism and cynicism are surrounded by the job seekers' suspension of the credibility of CEP information. As previously analysed, attracting candidates in recruitment is indeed a signalling process between the employer or recruiter as the sender and job seekers as the receivers, whereby the sender strives to signal the receiver about the sender's underlying value. The "Handicap" principle of signalling theory posits that signals must be costly or hard to fake to be perceived as honest. Besides a company's website and interviewer, the employer could adopt other methods to communicate CEP information; for example, positive word-of-mouth (especially, transferred from extant employees to their friends), information published by third parties (e.g., a company's achievements/awards – such as an ISO certificate, media and broadcasting) and employee testimonials [52].

Signalling should avoid *greenwashing* [32] and emphasise what would be most expected by the targeted candidates. In this regard, the five signal-based mechanisms work as predictors of applicant attraction. Depending on the type of candidates, a recruiter should detect what type of signal the job seeker is more attracted to and is more likely to interpret successfully. For example, graduate job seekers might have a tendency to pursue a job with organisations that are prestigious to enhance their profile and career mobility; generation Y might be more attracted to firms with ecological values; managers might be more attracted to an organisation with proactive strategies and workers whose jobs are primarily located in hazardous or polluted areas might care more about the working condition and green compensation. Personality and ecological tests may reveal whether the applicants are altruistic or egoistic so that interviewers can signal in different manners. Similarly, different kinds of CEP signalling should be adopted for different job positions.

Another principle of signalling theory is that the sender must have underlying quality; otherwise, false and misleading signals would be ignored by the receiver [40]. In recruitment, it is critical to avoid delivering fake signals because they could ruin the organisational prestige that takes years to regain. To avoid the hazards of *greenwashing*, companies are encouraged to adopt a proactive approach to CEP that requires them to engage in environmental strategies and practices (EMP and EOP) prior to the occurrence of negative incidents involving the firm; this approach engenders stronger trust and facilitates the attributions of the credibility of CEP information [38]. Finally, signalling effectiveness can be enhanced by signal frequency and signal consistency [40]. The recruitment plan of each position should be designed carefully before implementation to make sure that the signals transferred by various senders or recruiters are consistent in nature and that they simultaneously aim at activating at least one of the three predictors of applicant attraction, depending on what attracts the group of candidates the most.

In this respect, the recruiter/interviewer can choose to transfer the message via inspirational quotes, company's banners, environmentally heart-touching images, event photos/behind-the-scene photos, offerings or even recruiter/interviewer's personal story, feeling and perception toward the employer's environmental issues, etc. For example, heart-touching images concerning environmental problems along with inspirational quotes may activate the individual's environmental consciousness, information on green training and development program/promotion opportunity may activate the egoistic individual, information on company's green improvement programme and rewards for eco-initiative may activate the individual's creativity and ambition to contribute a significant impact through his/her work, a photo of staff in company's recognition may activate the individual's anticipated pride of being affiliated with the organisation. Importantly, this signalling requires the recruiters/interviewers' sensitivity toward the applicant's characteristics and expectation, and their pro-environmental behaviours to comply with the "Handicap" principle. Accordingly, the recruiters/interviewers play a foremost role as an activator of job seeker's perceived organisational attractiveness for environment.

## 2.4 Conclusion

Given the importance of qualified employees with the context of demographic change and the shortage of human resources in some countries, recruitment plays a crucial role, regardless of economic growth or recession [54]. Successful recruitment facilitates later practices after organisational entry, such as training, retaining and involving employees. Due to expectations from multiple stakeholders and organisational socio-environmental consciousness, organisations are to embed ES strategies into their core business practices. Therefore, green recruitment constitutes a great proportion in the guarantee of environmental performance by recruiting pro-environmental candidates. Accordingly, attracting pro-environmental job seekers is vital, especially when organisations are in an "arms race" for obtaining a high-quality workforce.

In assuming CEP as the indicator of an environmentally responsible employer, this chapter is set out to explore how signalling on CEP has a positive effect on job seekers' perceptions of organisational attractiveness for the environment before and during green recruitment process. Our proposed conceptual model of "Handicap" principle shows that signalling on CEP – in the form of EMP and EOP – will generate job seekers' perceived organisational prestige/anticipated pride, perceived value fit or perceived favourable treatment, which will lead to their perception of organisational attractiveness for environment. This model highlights the aggregated CEP construct that demonstrates employee participation at both the strategic and operational level. A recruitment process that limits information transmission will reduce potential organisational attractiveness [54]. Hence, to catch job seekers' attention and sustain their interest to stay, signalling on CEP should be implemented during the stages of green recruitment, provided that it respects and complies with the "Handicap" principle and that the message carries the underlying quality of the employer. In this regard, the two new concepts – perceived signal honesty and perceived signal consistency – shed light on the principle for candidate attraction in order to make job seekers have positive perceptions toward organisation.

This chapter has both theoretical contributions and practical implications. Theoretically, our research helps fill the gap in GHRM in a manner that promotes the involvement of green recruitment in the guarantee of quantity and quality of green HR, as well as increases an organisational image as environmentally-responsible employer. Relatedly, the findings enrich the literature in green recruitment by illuminating the "Handicap" principle for candidate attraction. In addition, our research has extended previous GRHM literature in some ways. First, our model contributes to the three signal-based mechanisms introduced by Jones et al. [48] and extends their research in the way that two new concepts – perceived signal honesty and perceived signal consistency – give an insight into the attracting mechanisms and prove for the "Handicap" principle in green recruitment. Second, thanks to the conceptualisation on CEP by Trumpp et al. [39], we contend that implementing CEP – both EMP and EOP – is important for attracting pro-environmental job seekers, as it signals that the company is responsible not exclusively for its business operation and the community, but also for the company's extant employees and the potential ones. Such responsibility increases organisational prestige.

From a practical perspective, green recruitment is worth practising to attract better participants who are more sensitive to environmental aspects and who voluntarily participate in a company's CEP practices. Signalling information on CEP enables employers to promote their prestige, thus gaining society's trust, as well as attracting and retaining pro-environmental employees. Furthermore, different job seekers are attracted to different "carrots" and "sticks", interviewers/recruiters could therefore leverage each mechanism, adopting adequate information to signal targeted candidates in the way that activates their positive perception toward ecological values and organisational attractiveness for environment. To enhance the effectiveness of green recruitment, respecting and making use of the "Handicap" principle is needed. Finally, each individual is more likely to trust in a specific signalling method and the probability that signalling effect is successful depends upon whether the pro-environmental individual actually receives the signals. As such, utilizing various methods of communication to signal consistent information facilitates successful attraction in green recruitment.

Our findings suggest some directions for future research. Given that previous studies in green recruitment are implemented in developed countries, empirical studies are encouraged to test this conceptual model of "Handicap" principle and the 10 propositions internationally, leveraging the different effects of the five signal-based mechanisms in different contexts. Further research would be on the relationship between organisational attractiveness for environment and job acceptance behaviour, as well as on other moderators, such as financial performance, organisational sector, organisational internationalisation, job characteristics and perceived signal frequency.

## References

[1] Talya N. Bauer, Berrin Erdogan and Sully Taylor. Creating and Maintaining Environmentally Sustainable Organizations: Recruitment and Onboarding. In S. E. Jackson, D. S. Ones and S. Dilchert, editors, *Managing Human Resources for Environmental Sustainability*, Jossey-Bass/Wiley, New York, 2012.

[2] Katrina Rogers and Barclay Hudson. The Triple Bottom Line: The Synergies of Transformative Perceptions and Practices for Sustainability. *OD practitioner*, 43(4):3–9, 2011.

[3] Markus J. Milne and Rob Gray. W(h)ither Ecology? The Triple Bottom Line, the Global Reporting Initiative, and Corporate Sustainability Reporting. *Journal of Business Ethics*, 118(1):13–29, 2013.

[4] Wayne Norman and Chris MacDonald. Getting to the Bottom of "Triple Bottom Line". *Business Ethics Quarterly*, 14(2):243–262, 2004.

[5] Jesús Ángel del Brío, Esteban Fernández and Beatriz Junquera. Management and employee involvement in achieving an environmental action-based competitive advantage: an empirical study. *The International Journal of Human Resource Management*, 18(4):491–522, 2007.

[6] Daniel W. Greening and Daniel B. Turban. Corporate Social Performance As a Competitive Advantage in Attracting a Quality Workforce. *Business & Society*, 39(3):254–280, 2000.

[7] Stanley M. Gully, Jean M. Phillips, William G. Castellano, Kyongji Han and Andrea Kim. A Mediated Moderation Model of Recruiting Socially and Environmentally Responsible Job Applicants. *Personnel Psychology*, 66(4):935–973, 2013.

[8] Charbel José Chiappetta Jabbour and Fernando Césa Almada Santos. Relationships between human resource dimensions and environmental management in companies: proposal of a model. *Journal of Cleaner Production*, 16(1):51–58, 2008.

[9] Corinna Dögl and Dirk Holtbrügge. Corporate environmental responsibility, employer reputation and employee commitment: an empirical study in developed and emerging economies. *The International Journal of Human Resource Management*, 25(12):1739–1762, 2014.

[10] Gilles Grolleau, Naoufel Mzoughi and Sanja Pekovic. Green not (only) for profit: An empirical examination of the effect of environmental-related standards on employees' recruitment. *Resource and Energy Economics*, 34(1):74–92, 2012.

[11] Charbel José Chiappetta Jabbour and Ana Beatriz Lopes de Sousa Jabbour. Green Human Resource Management and Green Supply Chain Management: linking two emerging agendas. *Journal of Cleaner Production*, 112(3):1824–1833, 2016.

[12] Susan E. Jackson, Douglas W.S. Renwick, Charbel J.C. Jabbour and Michael Muller-Camen. State-of-the-Art and Future Directions for Green Human Resource Management: Introduction to the Special Issue. *German Journal of Human Resource Management*, 25(2):99–116, 2011.

[13] Douglas W.S. Renwick, Tom Redman and Stuart Maguire. Green Human Resource Management: A Review and Research Agenda. *International Journal of Management Reviews*, 15(1):1–14, 2013.

[14] Julie Haddock-Millar, Chandana Sanyal and Michael Müller-Camen. Green human resource management: a comparative qualitative case study of a United States multinational corporation. *The International Journal of Human Resource Management*, 27(2):192–211, 2016.

[15] Irina V. Kozlenkova, Stephen A. Samaha and Robert W. Palmatier. Resource-based theory in marketing. *Journal of the Academy of Marketing Science*, 42(1):1–21, 2014.

[16] Lara D. Zibarras and Phillipa Coan. HRM practices used to promote pro-environmental behavior: a UK survey. *The International Journal of Human Resource Management*, 25(16):2121–2142, 2015.

[17] Andrea Kim, Youngsang Kim, Kyongji Han, Susan E. Jackson and Robert E. Ployhart. Multilevel Influences on Voluntary Workplace Green Behavior: Individual Differences, Leader Behavior, and Coworker Advocacy. *Journal of Management*, 43(5):1335–1358, 2017.

[18] Jennifer L. Robertson and Julian Barling. Greening organizations through leaders' influence on employees' pro–environmental behaviors. *Journal of Organizational Behavior*, 34(2):176–194, 2013.

[19] Regina Lülfs and Rüdiger Hahn. Corporate Greening beyond Formal Programs, Initiatives, and Systems: A Conceptual Model for Voluntary

Pro−environmental Behavior of Employees. *European Management Review*, 10(2):83–98, 2013.

[20] Cristina E. Ciocirlan. Environmental Workplace Behaviors: Definition Matters. *Organization & Environment*, 30(1):51–70, 2017.

[21] Icek Ajzen. The theory of planned behavior. *Organizational Behavior and Human Decision Processes*, 50(2):179–211, 1991.

[22] Chia-Jung Chou. Hotels' environmental policies and employee personal environmental beliefs: Interactions and outcomes. *Tourism Management*, 40:436–446, 2014.

[23] Megan J. Bissing−Olson, Aarti Iyer, Kelly S. Fielding and Hannes Zacher. Relationships between daily affect and pro- environmental behavior at work: The moderating role of pro-environmental attitude. *Journal of Organizational Behavior*, 34(2):156–175, 2013.

[24] Jenny Dumont, Jie Shen and Xin Deng. Effects of Green HRM Practices on Employee Workplace Green Behavior: The Role of Psychological Green Climate and Employee Green Values. *Human Resource Management*, 56(4):613–627, 2017.

[25] Pascal Paillé, Yang Chen, Olivier Boiral and Jiafei Jin. The Impact of Human Resource Management on Environmental Performance: An Employee-Level Study. *Journal of Business Ethics*, 121(3):451–466, 2014.

[26] Marta Pinzone, Marco Guerci, Emanuele Lettieri and Tom Redman. Progressing in the change journey towards sustainability in healthcare: the role of 'Green' HRM. *Journal of Cleaner Production*, 122:201–2011, 2016.

[27] Suzanne Benn, Stephen T.T. Teo and Andrew Martin. Employee participation and engagement in working for the environment. *Personnel Review*, 44(4):492–510, 2015.

[28] Douglas W.S. Renwick, Charbel J.C. Jabbour, Michael Muller-Camen, Tom Redman and Adrian Wilkinson. Contemporary developments in Green (environmental) HRM scholarship. *The International Journal of Human Resource Management*, 27(2):114–128, 2016.

[29] Sebastián Bruque, José Moyano and Ronald Piccolo. OCB and external–internal social networks: effects on individual performance and adaptation to change. *The International Journal of Human Resource Management*, 27(1):1–22, 2016.

[30] Guiyao Tang, Yang Chen, Yuan Jiang, Pascal Paillé, Jin Jia. Green human resource management practices: scale development and validity. *Asia Pacific Journal of Human Resources*, 56(1):31–55, 2018.

[31] Denise M. Jepsen and Suzanne Grob. Sustainability in Recruitment and Selection: Building a Framework of Practices. *Journal of Education for Sustainable Development*, 9(2):160–178, 2015.

[32] John Milliman. Leading-Edge Green Human Resource Practices: Vital Components to Advancing Environmental Sustainability. *Environmental Quality Management*, 23(2):31–45, 2013.

[33] Krista L. Uggerslev, Neil E. Fassina and David Kraichy. Recruiting Through the Stages: A Meta−Analytic Test of Predictors of Applicant Attraction at Different Stages of the Recruiting Process. *Personnel Psychology*, 65(3):597–660, 2012.

[34] Stéphane Renaud, Lucie Morin and Anne Marie Fray. What most attracts potential candidates? Innovative perks, training, or ethics? *Career Development International*, 21(6):634–655, 2016.

[35] Daniel B. Turban and Daniel W. Greening. Corporate Social Performance And Organizational Attractiveness To Prospective Employees. *Academy of Management*, 40(3):658–672, 1997.

[36] Kristin B. Backhaus, Brett A. Stone and Karl Heiner. Exploring the Relationship Between Corporate Social Performance and Employer Attractiveness. *Business & Society*, 41(3):292–318, 2002.

[37] Yuan-Hui Tsai, Sheng-Wuu Joe, Chieh-Peng Lin and Rong Tsu Wang. Modeling Job Pursuit Intention: Moderating Mechanisms of Socio-Environmental Consciousness. *Journal of Business Ethics*, 125(2):287–298, 2014.

[38] Chelsea R. Willness and David A. Jones. Corporate environmental sustainability and employee recruitment: Leveraging "Green" business practices to attract talent. In Ann Hergatt Huffman and Stephanie R. Klein, editors, *Driving change with I-O psychology*, pages 231–250. Routledge, London, 2013.

[39] C. Trumpp, J. Endrikat, C. Zopf and E. Guenther. Definition, Conceptualization, and Measurement of Corporate Environmental Performance: A Critical Examination of a Multidimensional Construct. *Journal of Business Ethics*, 126(2):185–204, 2015.

[40] Brian L. Connelly, S. Trevis Certo, R. Duane Ireland and Christopher R. Reutzel. Signaling Theory: A Review and Assessment. *Journal of Management*, 37(1):39–67, 2011.

[41] Adrian Bangerter, Nicolas Roulin and Cornelius König. Personnel selection as a signaling game. *Journal of Applied Psychology*, 97(4):719–738, 2012.

[42] Ray Karasek III and Phil Bryant. Signaling theory: Past, present and future. *Academy of Strategic Management Journal*, 11(1):91–99, 2012.

[43] Anthony Celani and Parbudyal Singh. Signaling theory and applicant attraction outcomes. *Personnel Review*, 40(2):222–238, 2011.

[44] Kia Kashi and Connie Zheng. Extending Technology Acceptance Model to the E−recruitment Context in Iran. *International Journal of Selection and Assessment*, 21(1):121–129, 2013.

[45] Tara S. Behrend, Becca A. Baker and Lori Foster Thompson. Effects of Pro-Environmental Recruiting Messages: The Role of Organizational Reputation. *Journal of Business and Psychology*, 24(3):341–350, 2009.

[46] Annika Wilhelmy, Martin Kleinmann, Cornelius J. König, Klaus G. Melchers, Donald M. Truxillo. How and why do interviewers try to make impressions on applicants? A qualitative study. *Journal of Applied Psychology*, 101(3):313–332, 2016.

[47] Annika Wilhelmy, Martin Kleinmann, Klaus G. Melchers and Martin Götz. Selling and Smooth-Talking: Effects of Interviewer Impression Management from a Signaling Perspective. *Frontiers in Psychology*, 8:740, 2017.

[48] David A. Jones, Chelsea R. Willness and Sarah Madey. Why Are Job Seekers Attracted by Corporate Social Performance? Experimental and Field Tests of Three Signal-Based Mechanisms. *Academy of Management Journal*, 57(2):383–404, 2014.

[49] Jos Bartels, Ad Pruyn, Menno De Jong and Inge Joustra. Multiple organizational identification levels and the impact of perceived external prestige and communication climate. *Journal of Oraganizational Behavior*, 28(2):173–190, 2007.

[50] Rosamaria Moura-Leite and Robert Padgett. The effect of corporate social actions on organizational reputation. *Management Research Review*, 37(2):167–185, 2014.

[51] Jin Feng Uen, David Ahlstrom, Shuyuan Chen and Julie Liu. Employer brand management, organizational prestige and employees' word−of−mouth referrals in Taiwan. *Asia Pacific Journal of Human Resources*, 53(1):104–123, 2015.

[52] David A. Jones, Chelsea R. Willness and Sarah Madey. Why Are Job Seekers Attracted by Corporate Social Performance? Experimental and Field Tests of Three Signal-Based Mechanisms. *Academy of Management Journal*, 57(2):383–404, 2016.

[53] Nancy Da Silva, Jennifer Hutcheson and Gregory D. Wahl. Organizational Strategy and Employee Outcomes: A Person–Organization Fit Perspective. *The Journal of Psychology*, 144(2):145–161, 2010.

[54] Denis Morin, Pascal Paillé and Anne Reymond. L'attraction organisationnelle: Une recension de la documentation scientifique. In Pascal Paillé, editor, *La fidélisation des resources humaines*, pp. 29–102. Les Presses de l'Université Laval, Québec, 2011.

[55] Kang Yang Trevor Yu. Person–organization fit effects on organizational attraction: A test of an expectations-based model. *Organizational Behavior and Human Decision Processes*, 124(1):75–94, 2014.

[56] Jeffrey Pfeffer and Gerald R. Salancik. The external control of organizations: A resource dependence perspective. Harper and Row, New York, 1978.

# 3

---

# Sustainable HRM: How SMEs Deal with It?

---

**Ana Luisa Silva and Carolina Feliciana Machado**[*]

School of Economics and Management, University of Minho, Portugal
E-mail: analuisalobarinhas@gmail.com; carolina@eeg.uminho.pt
[*]Corresponding Author

The purpose of this chapter is to reflect on the concept of sustainability. Not being a new topic, it has, however, gained considerable space in matters involving organizations. We want to reflect on its origin, the various definitions that exist about it, and the models that are associated with them. If the issue of sustainability involves organizations, it is imperative to reflect about the challenges in human resources management field, as well as its role for the organizations' sustainability. Considering that Portuguese business environment is deeply characterized by small and medium-sized enterprises (SMEs) we want to highlight the literature contributions about the way how sustainability is managed in SMEs.

## 3.1 Introduction

Organizations are currently dealing with a new challenge such as the association between the concepts of sustainability and human resource management (HRM).

"In the last decades, the issue of sustainability has also permeated discussions within organizations, with the understanding that business survival will depend on the re-dimensioning of the logic of economic thinking to a planning that considers social and environmental aspects as components of development and preservation of human existence" ([1], p. 3).

Organizations, today, face the need of rethinking their policies and practices. According to Rosa, Almeida, Dias and Junior [2], social and environmental problems resulting from unplanned actions require now new ways for companies deal with people and the environment. It has been asked to human resources management function to promote a joint work of individuals and organizations.

In this sense, the focus on this theme is due to the fact that this is a deeply discussed topic nowadays, in a growing development, that brings changes to the organizations and consequently to the way to manage their human resources.

The present chapter in addition to the introduction and conclusion is organized into three sections: the first section explores the concept of sustainability, where we think about the emergence of the theme, the various definitions and models that are adjacent to it. In the second section, we present some of the existing theoretical contributions about the role of human resources management to the organizations sustainability. Finally, in the third section, we present the perspective of some authors about the way sustainability is managed in SMEs.

## 3.2 Sustainability

According to Costa Lima and Montiveller (quoted by [3]), the theme of sustainability is not new. It was in 1972 at the Stockholm Conference that the concept was approached for the first time, contributing to the concept of eco-development. Its author was Ignacy Sachs, who believed that a country evolved if it gives particular attention not only to the economic area but also considered to the environmental and social field. In the 1980s, the concept of eco-development was replaced by sustainable development.

Sachs, who was one of the most important authors focusing the concepts of sustainability and eco-development, argued that sustainable development combines social, economic and environmental factors with the aim of respecting future generations. In its course, the concept of sustainability became more comprehensive, in order to incorporate new ideas ([4], p. 112):

"*social* dimension proposes social homogeneity, quality of life and social equality;

*cultural* dimension suggests balance, tradition, innovation and a combination between trust and openness to the world;

*ecological* dimension proposes the preservation of natural capital and the limitation in the use of these resources;

*environmental* dimension encompasses respect for natural ecosystems;

*territorial* approach deals with the balance between urban and rural configurations, the improvement of the urban environment and the development strategies of regions;

the *economic* approach addresses the economic balance between sectors;

*national policy* involves democracy and human rights and

*international policy* deals with the promotion of peace and international cooperation, international financial control, the management of natural and cultural diversity, and scientific and technological cooperation".

Nowadays, many organizations want to adjust their policies and practices to an issue that is on the agenda and in constant development, as it is sustainability, but it is not certain that everyone understands what the concept really means.

According to Rosa, Almeida, Dias and Junior [2], although the theme of sustainability is much talked about today, not everyone really knows what it refers to. Most of the time it is only associated with environmental issues, greatly reducing the definition of this concept. Speaking about sustainability is much broader than preserving natural resources.

"Sustainability means different things to different people. In essence, it conveys to meet the needs of the people today without compromising the ability of future generations to meet their own needs" (World Business Council for Sustainable Development, 2005, quoted by [5], p. 75).

We can define sustainability as the way to "operate the organization, without causing damage to living beings and without destroying the environment. By restoring and enriching it by identifying stakeholders, establishing open relationships with a search for common benefits, thus generating in the long run, more profit for the organization as well as more social, economic and environmental prosperity for society" (SAVITZ referred to by [6], p. 197).

Another definition states that sustainability "constitutes the capacity to maintain several social systems functioning through objective actions aimed at solving environmental crises and promoting sustainable development. This development occurs, therefore, through the application of sustainability and is evidenced in the interaction among natural, social, cultural and economic interests, ratifying the fact that the term sustainability is not only dedicated to environmental or ecological domains" ([4], p. 111).

Jabbour and Santos consider organizational sustainability as "management practices that aim at non-temporal vitality of the organization, favoring criteria of economic, social and environmental performance based on ethics and transparency. To the authors, sustainability requires investment, at the same time that adds value to the organization and shareholders, stimulating a more sustainable world" ([6], p. 197).

Claro, Claro and Amâncio believe that "there is a lack of consensus about its meaning, but the different definitions present as a common point that sustainability is composed of three dimensions that related itself: economic, environmental and social" ([6], p. 197).

The idea of sustainability in the organizational context has changed with the creation of the Triple Bottom Line model. Its author, John Elkington (referred by [7]), believes that the social responsibility of organizations can be the way to overcome environmental and social crises.

According to Elkington, "to the organization, the pursuit of profit and the maximization of its market value is recognized by the economic dimension, but when it comes to sustainable development, the traditional concept of accounting profit is not enough. Physical, financial, and human capital, which forms the economic capital, need to be included in this discussion with environmental and social issues" ([7], p. 76).

The author talks about the importance of the organization to find a balance between three dimensions: economic, social and environmental and thus achieve success.

In order to better understand this model, it will be necessary to analyze each dimension individually. According to Rosa, Almeida, Dias and Junior [2], the economic dimension concerns the formal economy and informal activities. In turn, social dimension refers to the qualities of individuals, what characterizes them, both within the organization and in the external environment. Finally, the environmental dimension encourages organizations to rethink their practices in a way that protects the environment, thereby implementing environmental concerns in the routine of companies.

In sum, what the author of the model intends to convey to the organizations is that it is important that they "evaluate their results not only by reference to financial performance but also to their impact on the broader economy, on the environment and society where they act" ([6], p. 197).

In order to clarify what is essential for an organization to be considered sustainable, Léon-Sorian, Munoz-Torees and Chalmeta-Rosalen consider that "sustainable organization would be the one that can effectively generate profits for owners and shareholders, protects environment and improves the lives of the people with whom it interacts" (cited by [8], p. 431).

In short, a sustainable organization has to integrate economic, environmental and social practices. In economic sustainability, "economic viability is the central argument for sustainable development, since it is through the circulation of wealth and profit generation that jobs are provided and is given to the community the possibility of improving its living conditions" (Autio, Kenney, Mustar et al., cited by [8], p. 431).

On the other hand, environmental sustainability concerns the "rational use of natural resources, such as energy and materials, as well as the preservation and recomposition of natural spaces" (Krajnc & GLAVIC, cited by [8], p. 431).

Finally, social sustainability refers to "stimulating equality and the participation of all social groups in the construction and maintenance of the balance of the system, sharing rights and responsibilities" (GreenWood, cited by [8], p. 431).

After this reflection on sustainability and sustainable organizations, it is important to emphasize that today it is not only required for organizations to focus on environmental issues and to preserve resources, and it is necessary to align sustainability practices with the way people are managed. Human resources management is then responsible for raising employees' awareness of this issue and fostering attitudes towards organizational sustainability.

## 3.3 Sustainable HRM

It is noted that human resources also begin to have concerns about organizational sustainability. According to Penha Rebouças, Abreu e Parente [9], the Society for Human Resource Management analyzed in 2011 the human resources practices to keep up with the economic and employment challenges of today's world, and concluded that one of the concerns is to play a role in the development of sustainability.

According to Paauwe (referred to by [10]) with the advances of technology and, consequently, the society, human resources management has to evolve and renew itself. Evolution brings with it new problems and new needs, people have new desires and goals, and organizations have to seek to respond to these new demands and adapt to their environment in order to be sustainable. Thus, the organization should pay attention to what people want, when it establishes its policies and practices, and thus manage in a sustainable way.

This makes it urgent "to deepen the debate about the interaction between the area of people management and sustainability in organizations, because within this new perspective, people management professionals must also act to respond to global pressures, as social, technological, environmental, political, economic and demographic changes" (Ulrich, cited by [4], p. 109).

In view of the society evolution, as focused earlier, and according to Aligleri, Aligleri and Kruglianskas (quoted by [3]), human resources management needs to develop some attitudes of commitment to employees, such as: accepting diversity in the organization; not have discriminatory attitudes in recruitment and selection; not allowing harassment; promote the physical and mental well-being of employees; have salary equality policies between men and women; and not dismiss employees based on criteria such as age or the civil status.

Dhamija [5] argues that sustainable human resource management evolves through four phases: *conception; personal department; strategic management of human resources and sustainable human resource management.*

The *conception phase* refers to the birth of organizations, in which existing practices and policies are not yet formalized considering the restrictions of the available budget. The *personnel department phase* refers to the operationalization of activities. The *strategic human resources management phase* concerns to the alignment of human resources practices with the organization's objectives. Finally, in the *phase of sustainable human resources management,* sustainable HRM has as mission to align its objectives with those of the organization, as well as to work towards practices of diversity management and environmental management.

This way, and according to Monteiro, D'Amorim, Machado and Zago [1], sustainable performance of human resources should define policies and implement practices in organizations that respond to this end.

In this sense, in their study, they aimed to know the strategic actions of the human resources area that contribute to the preservation of organizational sustainability. They were based on the sustainability models suggested by the Global Reporting Initiative (GRI), as well as, on the contribution of other authors, and gathered a set of practices that favor sustainable human resource management, namely:

(a) *Employees training* – according to Colbert and Kurucz (referred to by [1]), the human resources manager' function should focus on create a balance between people and decisions about sustainability, hiring collaborators that fit this profile, sensitizing them to sustainable practices;

(b) *Gender, age and minority diversity* – according to GRI and Madruga and Fagan (referred by [1]), organizations must always respect the context where they are inserted, living together and accepting the difference;

(c) *Native employees* – according to GRI (referred by [1]), organizations must bet on the hiring of native workers, since the fact that they are better acquainted with the zone, can have a direct influence in its development;

(d) *Social responsibility of the company (tasks and salaries, benefits and rewards)* – according to the GRI (referred by [1]), organizations must pay their employees' salaries that match with the market as a whole, maintaining equal pay between men and women;

(e) *Rewards to workers* – according to Gonvindarajulu and Daily (quoted by [1]), rewards motivate employees to opt for sustainable practices;

(f) *Performance appraisal* – according to the GRI (referred to by [1]), this instrument, when built with coherent criteria and knowledge of all, is a way of differentiating and recognizing the employees. It is suggested that "within the performance appraisal should exist a session that evaluates the employee's performance in activities that drive sustainability within companies" (Jabbour and Santos, cited by [1], p. 6);

(g) *Career management* – Iles (referred by [1]) suggests that the employees career should be developed in a sustainable manner, distinguishing and retaining employees who show practices that go beyond that which is required by contract;

(h) *Training* – Fernandéz, Junqueira and Ordiz (cited by [1], p. 6) argue that "all training and development of companies must be guided by the vision of sustainability and should be taught to all employees of the company, aiming to incorporate the values and skills inherent to actions that sustain organizational sustainability".

The authors also highlight other elements that human resources management must take into account in order to have a sustainable organization, such as, concern for workers' quality of life and their mental health, health and safety actions, organizational culture and leadership.

Human resources management can play an important role in the environment management through three dimensions (Wehrmyer, cited by [11], p. 40):

1. "to support the environmental management system through training, communication and employee motivation;
2. organizational changes, incorporating the environmental variable into the values of the organization, developing competences focused on environmental management and acting ethically on issues related to the environment and
3. insert the environmental variable in the human resources practices of recruitment and selection, performance appraisal, compensation and training".

According to Freitas, Jabbour and Gomes [11] in order that HRM can effectively participate in the organization sustainable management, human resource managers must play an active role in organizations, knowing its culture and values, the organization' environment, the diverse practices (recruitment and selection, training, remuneration, benefits and performance appraisal, among others) and the objectives that the organization intends to achieve. It is up to them to be proactive, creating and increasing new techniques that aim at a sustainable organization in a constantly changing environment.

Considering the relevance of the sustainability concept and the environmental issues, and according to Freitas, Jabbour and Gomes [11], in 2009, at the Academy of Management Annual Meeting, Professor Susan Jackson – a researcher in the human resources management field – established the Green Human Resource Management (Green HRM) international forum, which demonstrates the need to add the environmental dimension to human resource management.

With the involvement of human resources management in sustainability practices, employees are expected to be more proactive. To Camargo, Liboni and Oliveira [7] with this more proactive attitude, the employees began to reflect on their role within the organization and think about sustainable practices and plans, appearing, this way, the term of Green HRM.

This way, "giving voice to these employees and encouraging them to participate in decision making and problem solving will ensure empowerment (Ali and Ahmad 2009; Matthews et al. 2003). When they are empowered in pursuing green tasks of organization this will be termed as 'Green Employee Empowerment' which will come under the vast umbrella of green HR" ([12], p. 238).

In summary, the design of Green HRM with the active participation of the collaborators has gained a lot of relevance since people are the central element of the organizations.

## 3.4 Sustainability in SMEs

In Portugal, most of the companies are small and medium-sized enterprises (99.9%), being about 95.7% of them considered micro-enterprises [13].

Through these results, we can conclude that SMEs have a great representation in the Portuguese business field.

According to Silva, Albuquerque and Leite [6], SMEs are responsible for the generation of employment and employability of many people, which makes it pertinent to analyze how human resources management develops in these organizations, and, in turn, that these organizations become sensible to sustainability issues.

To Leite, Santos and Oliveira [14], facing the need to think about environmental issues, SMEs are adopting environmental management practices, being their managers aware of and realizing the benefits of a sustainable management, such as a recognition by society, more satisfied customers, the possibility of conquering new markets and more likely to obtain financing and aid from governments.

Regarding the implementation of sustainability practices in SMEs, Collins et al. (referred by [6]) point out that small companies have greater difficulty since they lack experience and have lower capital in relation to large companies; however, SMEs are very influenced by the environment in which they are inserted and where environmental management practices have power. Small and medium-sized enterprises are, this way challenged, to find the means to take sustainable policies and practices adopted by large enterprises into their own context.

According to Baumann-Pauly (quoted by [15], s/d), in studies that aimed to understand if SMEs have better or worse instruments in comparison with large organizations to deal with issues related to sustainability, it was not possible to arrive to a consensus.

However, there were identified some constraints faced by SMEs when they adopt sustainability practices. "The first is the perception that their individual environmental impact is small; the second is the lack of expertise and understanding of strategies to respond to environmental issues; and finally, as a third barrier, there is a cost for adopting proactive environmental behavior" (Collins et al., cited by [6], p. 207).

According to Silva, Albuquerque and Leite [6], studies carried out in small, medium-sized and large companies in New Zealand concluded that having a network can be an advantage for the adoption of sustainability practices in SMEs because they can obtain several apprenticeships with other organizations.

## 3.5 Final Remarks

The analysis of the theoretical contributions shows that the theme of sustainability is emerging and in the current market, organizations cannot be indifferent to thinking about practices that lead to sustainable management.

Human resources manager is called to intervene in the search of organizational sustainability. Human resources management has to be strategically thought by creating policies and practices that are integrated with the organization's strategies and objectives. Human resources management must prioritize people and play a key role in the organization success with a focus on environmental practices.

Thus, human resource management must be an ally of the organization in seeking good results, not neglecting the social context and always taking into account sustainable development.

As we have seen, small and medium-sized enterprises are not left out of this reality and are also challenged to seek mechanisms that allow them to have practices that contemplate sustainability.

## References

[1] Monteiro, M., D'amorim, A., Machado, A., and Zago, C. (2010). *Sustentabilidade: ações estratégicas da área de Recursos Humanos em organizações sustentáveis.* XXX Encontro Nacional de Engenharia de Produção. Maturidade e desafios da engenharia de produção, competitividade das empresas, condições de trabalho, meio ambiente. 12–15 October. São Carlos, SP, Brasil, 1–15.

[2] Rosa, C., Almeida, M., Dias, V.G., and Junior E. (2012). Gestão de pessoas e a sustentabilidade. *Anais da VI Mostra Científica do Cesuca 1(6),* s/p.

[3] Alvares, K. and de Souz, I. (2016). Sustentabilidade na Gestão de Pessoas: práticas e contribuições às organizações. *Revista Gestão Organizacional, 9*(2), 24–38.

[4] Oliveira, J.; Estivalete, F.; Andrade, T. and Costa, V. (2017). Gestão de pessoas e sustentabilidade: construindo caminhos por meio das práticas de capacitação. *Brazilian Journal of Management/Revista de Administração da UFSM, 10,* 108–126.

[5] Dhamija, P. (2013). Human resource management: an effective mechanism for long term sustainability. *The Clarion, 2*(1), 74–80.

[6] Silva, M.; de Albuquerque, L. and Leite, N. (2013). O alinhamento entre a estratégia de Gestão de Pessoas e sustentabilidade em uma pequena empresa: AMAZONGREEN. *Revista de Ciências Empresariais da UNIPAR, 13*(2), 193–216.

[7] Camargo, J.; Liboni, L. and Oliveira, J. (2015). Gestão ambiental de recursos humanos e nível de envolvimento de colaboradores nas organizações. *RAM. Revista de Administração Mackenzie, 16*(2), 72–91.

[8] Kuzma, E.; Doliveira, S. and Silva, A. (2017). Competências para a sustentabilidade organizacional: uma revisão sistemática. *Cadernos EBAPE. BR*, *15*, 428–444.

[9] Penha, E.; Rebouças, S.; Abreu, M. and Parente, T. (2016). Percepção de responsabilidade social e satisfação no trabalho: um estudo em empresas brasileiras. *REGE-Revista de Gestão*, *23*(4), 306–315.

[10] Freitas, W.; Jabbour, C. and Santos, F. (2009). Rumo à sustentabilidade organizacional: uma sistematização sobre o passado, o presente e o futuro da gestão de recursos humanos. *II Encontro de Gestão de Pessoas e Relações de Trabalho*, 1–15.

[11] Freitas, W.; Jabbour, C. and Gomes, A. (2011). Gestão ambiental: um novo desafio para os profissionais de recursos humanos?, *Revista Cesumar-Ciências Humanas e Sociais Aplicada*, *16(1)*, 29–47.

[12] Tariq, S.; Jan, F. and Ahmad, M. (2016). Green employee empowerment: a systematic literature review on state-of-art in green human resource management. *Quality & Quantity*, *50*(1), 237–269.

[13] INE, (2014). Empresas de Portugal 2012. Lisboa: INE. Retrieved from https://www.ine.pt/xportal/xmain?xpid=INE&xpgid=ine_publicacoes& PUBLICACOESpub_boui=210758098&PUBLICACOESmodo=2, accessed in May 2018.

[14] Leite, K.; Santos, M. and Oliveira J. (2011). Sustentabilidade: Fator Preponderante nas Micro e Pequenas Empresas. *Revista Administração Eletrônica*, (6), 1–10.

[15] Canto, N.; Rodrigues, R. and Alves, A. (s/d). *Práticas de Sustentabilidade em uma Média Empresa: um estudo de caso em uma indústria gaúcha do setor de alimentos. ENGEMA* - Encontro Internacional sobre Gestão Empresarial e meio ambiente. Retrieved from http://www.engema.org.br/XVIENGEMA/305.pdf, accessed in May 2018.

# 4

# The (Un)sustainable Process of Devolution of HRM Responsibilities to Line Managers

**João Leite Ribeiro* and Delfina Gomes**

School of Economics and Management, University of Minho, Portugal
E-mail: joser@eeg.uminho.pt; dgomes@eeg.uminho.pt
*Corresponding Author

Human Resources Management can play an important role in the sustainability of organizations and it has over the years assumed this role, although not always well understood at various hierarchical and functional levels. The central objective of this study is to understand the perceptions and reactions of different organizational actors about the process of devolution of Human Resources Management responsibilities to line managers. Understanding how the devolution of responsibilities to the sphere of assignments of line managers is perceived is relevant to the sustainability of this process of reassignment of responsibilities within organizations. In this study, 257 interviews were conducted with organizational actors, and Grounded Theory was applied to analyse the data. The interviewees were employed at 10 companies (3 multinational and 7 Portuguese companies) and held different hierarchical positions: top managers, human resources managers and collaborators, peers of human resources managers, and collaborators from different organizational functions. The study highlights that it is important that organizations and their leaders acknowledge that although the development of human resources practices is the responsibility of the human resources department, the implementation of those practices is the responsibility of line managers. At the same time, in the same organization, it is possible to find arguments in favor and against the devolution, and even change of opinions and behaviors according to personal strategies and tactics.

At the level of contributions to the practice, knowledge of perceptions can be an asset for adjustments between professionals, as well as to reduce role and responsibility ambiguities, enhancing attitudes and more sustainable behaviors.

## 4.1 Introduction

The quality of the management function can be an important factor for the sustainability of organizations, by contributing to their effective development or even ensuring their survival in more hostile environments [13, 98, 118, 119, 122, 123, 130, 132, 147]. Ensuring sustainable development should be one of the main responsibilities of a manager. For this to happen, in addition to having a sense of the vision and mission of the organization to which the manager belongs, he must know the objectives, the resources available to achieve them and have the appropriate hard and soft competencies [12, 13, 98, 132].

Human Resources Management (HRM) can play an important role in the sustainability of organizations and, in many cases, it has over the years assumed this role, although not always well understood at various hierarchical and functional levels [9, 94, 95, 125, 132, 134, 135]. Also part of the historical characterization of this area of management is the gap between narrative, discourse and rhetoric about the importance of human resources and the way in which their management is often carried out [132].

The central objective of this study is to understand the perceptions and reactions of different organizational actors about the process of devolution of HRM responsibilities to line managers. Thus, this study addresses the following research question: How are line managers perceived by different organisational actors as intervenient in HRM and in what way are they seen as responsible for the management of the people and teams they coordinate? It is important to realize the perception about the role these managers have of HRM insofar as their intervention can be more or less consistent with the adoption of more sustained perspectives. To understand how the devolution of responsibilities of this area to the sphere of assignments of line managers is perceived is relevant to the sustainability of this process of reassignment of responsibilities within organizations.

Sustainability is understood here as the result of management behaviors which contribute to ensure organizational balances and increase consistency at horizontal and vertical fits, which HRM seeks to have in different contexts of action and according to defined guidelines [9, 132, 147].

This study contributes to deepen the knowledge and understanding of the perception and reaction assumed by different organizational actors in the process of devolution of HRM responsibilities and activities to the line managers. As argued by Sikora and Ferris [132], "while in most firms, the human resources department is responsible for the development of effective HR practices, the implementation of those practices ultimately falls to the firm's line managers" (see also [9]). At the level of contributions to the practice, knowledge of perceptions can be an asset for adjustments between professionals, for the development and acquisition of hard and soft HRM skills, as well as to reduce role and responsibility ambiguities, enhancing attitudes and more sustainable behaviors [9].

The remainder of this chapter is structured as follows. The next section presents a review of the literature beginning with the contextualization of management and HRM advances and retreats, and the historical evolution and scope of HRM. The methodology of the study and a description of the participants are presented in the subsequent section. Next, different organizational actors' perceptions on how line managers intervene in HRM and how are they seen as responsible for the management of the people and teams they coordinate are described and analysed. The final section offers a discussion and conclusion.

## 4.2 Management and HRM: Contexts of Advances and Retreats

Given that the central objective of this study is to understand the perceptions and reactions of different organizational actors about the process of devolution of HRM responsibilities to line managers, it is pertinent to contextualize the advances and setbacks of management and HRM. Strategic decisions, operational and administrative tactics that the top management of an organization or formal HRM managers assume can significantly contribute to defining and/or delimiting decision-making at the level of delegating responsibilities to line managers [9, 90, 132, 147]. Therefore, it is important to have knowledge of the advances and retreats of management and HRM.

The first approaches to management were characterized by being excessively normative, controlling, based on very strict rules and procedures, with the concern focused on increasing levels of income and productivity. Gradually, throughout the 20th century there has been an evolution towards contingent and complex approaches, in which the paradigm has become that of competitiveness and quality [9, 10, 12, 13, 24, 35, 44, 82, 94, 105, 106, 125, 126, 144].

Although with advances and retreats, management has gradually ceased to be seen exclusively as a set of techniques and practices that are perfected to assume strategic, operationally sound values, principles and institutional processes. Management has thus concomitantly assumed to be a social practice, more systematic in terms of concrete actions and developments visible in an emergence of concepts and actions linked to ethical and social responsibility [12, 98, 123, 130, 132].

The theoretical and practical developments of HRM occurred mainly from the 1990s, with organizational structures characterized by greater fluidity, flexibility and decentralization [51, 56, 57, 61, 147, 156]. Greater emphasis is placed on the formation and development of multifunctional and multidisciplinary team networks, revealing a new mindset in managing and dealing with situations [5, 13, 53]. The predominant organizational culture is based on the principles of MacGregor's Y theory, oriented towards the future and valuing innovation, knowledge and creativity [61, 81, 102, 111, 123, 129, 132, 149–151]. The organizational environment is characterized by high mutability and unpredictability, leading to intense and discontinuous changes [5, 100, 117, 119, 133].

People are perceived as proactive, gifted with intelligence, with differentiated knowledge, skills and abilities that must be identified, developed and empowered in terms of organizational performance and individual motivation. Freedom and autonomy, as well as commitment and responsibility, emerge as the main motivational strategies of workers [61, 75, 141, 157]. The underlying conception assumes the worker not only as a human being motivated by economic needs, but also emphasizes the importance of social, psychological, and emotional factors. It also highlights the relevance of informal structures and the motivational and leadership processes, as conditioners of workers' performance and commitment [12, 13, 16, 29, 50, 52, 122, 157]. It is in this perspective that the transversality of management and HRM and the role that different actors can develop in the act of managing people and teams must be considered and perceived.

The last two decades of the 20th century correspond to two levels of management, respectively, the information and digital phases, and it is not easy to define the point in time that separates, at least in theoretical terms, these two phases [130]. The management concepts, associated to these two moments, are characterized by: relevance of new values; emergence of new ways of communicating; virtualization of human relations; increasing and absorbing proliferation of social networks; exponential development of technology; increasing dematerialization of work and its computer programming; emergence of new work behaviors; and by the emergence of new professions. Hence emerges new challenges for management, at this time globally designated as the *era of knowledge* [5, 14, 19, 26, 27, 78, 114, 127, 129, 130, 142].

These new challenges involve the (re) understanding of the concept of work, the new type of workers, and the relationships between different generations of workers with diverse values, interests, needs and perceptions of the world [9, 37, 114, 122, 125, 130]. These aspects will imply a reinvention of the ways of managing, with emphasis on innovation on the art/science of managing [29, 30, 61]. They also imply new leadership and new ways of leading people in multicultural, diversified environments with equally differentiated states of mind and capable of transforming the organization into a learning organization [9, 13, 37, 50, 117, 122, 125, 129, 132, 144, 157].

A set of new trends emerged in the 21st century placing emphasis on organizational performance in general and consequently on the contributions of HRM in particular. Increased productivity and pressure for competitiveness and quality of management are some of these trends [5, 13, 61, 87, 157]. At the organizational level, there is a tendency for the discourse to focus on strategic management, be it operations, research and development, information technology, marketing plans, innovation and creativity, logistics processes, administrative systems, accounting and financial policies and practices, and of HR [61]. At the academic level, HRM research highlights the increase in a critical and assertive role of this area of management in organizations [33, 81, 86, 87, 154]. It is emphasized that effective HR management requires an understanding of the evolutions and trends that emerge in a complex and volatile world and the redefinition of the orientation to make HRM present, effective and capable of being a distinctive element for management and for the company [13, 132, 157]. Thus, HRM must adopt a strategy of diagnosis and anticipation of reality and have an excellent ability to read events in the contexts in which they arise, adopting, whenever possible, a proactive approach [5, 13, 61, 63, 157, 160, 162, 163].

## 4.3 The Evolution of HRM

The historical evolution of HRM reflects, above all, the changes in the ways of conceiving and managing people. It goes from a conception of people as mere static factors of production, to a conception of people perceived as proactive human beings with skills and capacities that must be encouraged and developed [5, 16, 63, 160].

Management concepts have been evolving since the beginning of the 20th century, marked by absolute rationality, for more recent conceptualizations based on principles of contingency and complexity, with inevitable reflections on HRM [5, 31, 51, 140]. As stated by Cabral-Cardoso [29], with regard to HRM, it is "by nature contingent, which means that its practices must be adequate and adjusted to the circumstances and that only a detailed analysis of the surrounding conditions enables the development of a policy and to establish lines of action at the level of the group or company".

In terms of personal function, the first phase is in the first three decades of the last century. At that time, the first industrial production units were created, characterized by an increasing control by the bosses in relation to the workers. The aim was increasing profitability of the industrial units with the implementation of effective systems of planning, organization and division of functional responsibilities [4, 29, 34, 41, 95, 96, 113, 135]. Several paradigms have emerged through the 20th century and, since the 1970s to the present time, there was the emergence of a new HRM model that translates the paradigm shift, with the shift from the logic of productivity to the logic of competitiveness and quality.

This is the most recent phase of HRM and is called strategic human resources management (SHRM). At this stage greater importance is attributed to satisfaction and motivation at work and there is a position of equality of the personnel function compared to other organizational functions, a situation for which greater theoretical consistency has contributed [8, 9, 20, 56, 59, 63, 125, 156]. This strategic scope of HRM is also embedded in the development of HRM policies, techniques and practices. Some of the factors that have contributed to this sophistication of HRM are: the globalization process with all its consequences, namely, the increase in global competition; rapid developments in technology; the change in the concepts of work and industrial relations; the awakening of civil society and business to issues such as ethics and corporate social responsibility; sociocultural and demographic issues that bring new challenges and expectations for organizations stemming from entering in the labor market of young workers and their coexistence with other generations of workers; and, greater consolidation of the benchmark for competitiveness and quality and respective indicators of efficiency and effectiveness [9, 12, 13, 29, 34, 56, 57, 61, 63, 92, 108, 125, 147, 156].

Developments in labour law and economic growth drove the creation of personnel management services. These services, in addition to maintaining recruitment and selection, training, trade union negotiations and payment of salaries, began to value and develop performance appraisal processes, as well as labor-based planning systems [1, 2, 9, 12, 29, 39, 40, 92, 125, 144]. Workers are given a fundamental role in relation to the organization's strategy, being seen as people with intelligence, critical spirit, and active. These workers, if properly managed, add value to the organization and become a competitive and distinctive advantage [10, 13, 20, 29, 30, 43, 58, 59, 61, 70, 117–119, 134, 142, 143, 147–153].

This phase is initially characterized by some ambivalence towards the person, and there are situations where this is considered as any other business resource, which must be managed at the lowest possible cost. It assumes the assumptions of the HRM hard model. according to other perspectives the person is seen as an investment that can and must be empowered, in which people are perceived as proactive, intelligent beings capable of suggesting important changes in the form and content of their functions. People are also seen as strategic and advantageous resources for the organization. [1, 2, 9, 61, 89, 91, 125, 138, 146, 147].

This stage of HRM is strongly marked and influenced by the development of two key areas: strategic management and organizational behavior, which helps HRM to aspire to play a strategic role in organizations [1, 2, 9, 20, 21, 61, 89, 110, 116, 125, 147]. This strategic importance has been increasing in the last decades through a greater representation in the company's management bodies, a greater and more consistent approach to the business of organizations. Equally significant aspects of HRM's visibility and strategic nature are: enrichment of functions, emergence of new themes, increase of responsibilities at different organizational levels, attribution of greater power and influence, more widespread recognition of its relevance and importance, and a greater and more consistent functional identity [1, 2, 5, 9, 29, 61, 89, 110, 125, 138, 147].

The end of the 20th century is marked by economic growth, aggressive brand marketing, high technological development, emergence of new countries and new economies (e.g., China, India, Brazil, Russia), and serious demographic imbalances in different parts of the globe. But it is also marked by the emergence of new HRM themes, such as issues of equal opportunities, management of workforce diversity, work-life balance, and privacy issues. There is also a decrease in the impact and capacity of trade union influence.

All of these facts have had and have implications for HRM [5, 7, 13, 18, 20, 24, 37, 48, 91, 94, 122].

Already in this new century, other important events occurred. The 11 of September of 2001 caused a considerable impact, in particular in terms of insecurity and the following military retorts, the new oil shock and the crisis that settled in 2008, initially in the financial sector to rapidly expand into the economic sector in 2009 and 2010 with ideological, political and social repercussions [5, 6, 77, 88, 101, 112, 139]. This crisis has exposed structural weaknesses, deficits in values, turbulence in the job market with bankruptcies, lay-offs, relocations, disenfranchisement of leaderships and institutions, and an increase in social scourges. In particular, unemployment continues to rise very significantly, especially in some European Union countries, with inevitable and unpredictable consequences on the labour market, on labour relations, on social interactions and on HRM [13, 55, 61].

Organizations should be able to adapt their strategies to these new times, with policies and management practices that are both consistent and innovative and with added value for the organization in the specific contexts in which they operate [13, 55, 61, 62, 69, 131, 114, 157]. In this way, it can be stated that the HRM is organized and developed around four axes described by Peretti (1996) [116] and Brandão and Parente (1998) [20], which are still valid, as follows. First, *processes of flexibility* of the employment, the nature of the contractual tie, the duration and management of the time worked. Second, *systems of individualization*, namely of the management of the careers and of talents and professional paths. Third, *remuneration policies* and incentives linked to performance evaluation and the principles of meritocracy. Finally, *learning processes and social innovation*, as a need for permanent adaptation to the internal and external environment [5, 13, 20, 62, 68, 71, 114, 157].

## 4.4 Scope of HRM

The expression "human resources management" has given rise to a significant and controversial debate, especially in the Anglo-Saxon world. This controversy makes it difficult to delineate its boundaries, areas of responsibility, and hence a precise formulation about its meaning and impact [4, 5, 9, 32, 63, 79, 94, 125, 134–136, 147]. Before deepening the controversy over the differences between personnel management and HRM, some perspectives on the definition, scope and some of the HRM models are presented.

Bratton and Gold (1999 [21]; see also [24]) define the scope of HRM as the part of the management process that specializes in managing people in organizations. HRM emphasizes that employees are key to achieving competitive advantage. That HRM practices need to be integrated with company strategy and that HRM's experts assist management in meeting efficiency and equity goals.

Underlying this perspective is the fact that in HRM what really makes a difference is the nature of the resource in question, which is the person [5, 9, 42, 43, 59, 79–81, 116, 125, 147, 152, 153]. Compared to other organizational resources, people not only differ in themselves as they differ from each other. People differ from each other in terms of competencies and cognitive characteristics, practical skills and abilities, and personality traits, among others. They differ in their perceptions, whether physical or social, in their perception and appreciation of the different roles they play in society, in life experiences and even in levels of motivation and commitment (intra-individual and inter-individual variability). According to McShane (1995 [104]; see also [12, 63, 81], the human being is perceived as a resource with potential, creative and complex, whose behavior is influenced by a series of factors of the individual forum and the surrounding environment. For this author, the behavior and performance of the "human resource" is a function of at least four variables: competencies, motivation, role perception and situational contingencies (cf. [13, 157]).

In addition to the problems related to the definition and conceptualization of its key resource, HRM has been the object of attempts to systematize its responsibilities regarding the functional areas, particularly with regard to concepts, methods, techniques and practices. Therefore, according to Bratton and Gold (1999 [21]; see also [12]), HRM encompasses four functional areas. The first is staffing, that is, in the detection, selection and allocation of people with the skills, knowledge and experience appropriate to the exercise of the organization's functions. The second is the definition and management of rewards systems. The third concerns the development of employees that aims to analyse their training needs in order to fill them and make the organization more efficient and effective. The latter is related to employee retention, which consists of maintaining a competent workforce and meeting the defined criteria and procedures [29, 61, 162, 163].

For Brewster (1997) [23], since the beginning of the 1980s the personnel management function and its departments have been the subject of multiple debates at the level of the professional class itself in an attempt to seek a common professional identity. This author refers to the need for professionalization, qualification, and professionalism that the HRM area needs for several reasons [3, 11, 38, 46, 64, 66, 72, 76, 99, 159]. One of the reasons has to do with the fact that, at least at the discourse level, the human resource is considered by many managers as the most valuable resource of the organization and the rarest because it is the only one impossible to be copied. Thus, the human resource needs to be managed properly and professionally, that is, depending on the business of the organization and its evolutionary trends [5, 9, 42, 45, 49, 66, 111, 118, 119, 125, 147, 158]. Another reason has to do with the changes that have occurred in the HRM function and the new challenges that organizations face at this level, such as: meaning and emphasis given to HRM; characteristics and responsibilities of this area of management; and responsibilities of line managers [22, 23, 25, 28, 29, 67, 73, 74, 84, 90, 107, 120, 132].

The advantage of considering these three interconnected challenges has to do with, according to Brewster (1997), how the strategic importance of people is affecting the way line managers operate. Line managers are concerned with the nature of work and flexible working methods have implications for staff costs and affect all involved in strategic decisions see also [5, 9, 54, 73, 74, 83, 85, 115, 120, 125, 147, 161].

The HR body frames several of these themes and presents the following responsibilities: diagnosis and proactive action in face of the pressures for change; to promote the importance of HRM through actions that favor the creation of organizational value; to support dynamically and critically the situations that have to do with changes in the nature of work, particularly the temporal flexibility of work, contractual and functional flexibility; and finally the support needed by line managers [15, 63, 66]. While this list is, in general, an appropriate description of some of the responsibilities of the HRM function, these activities vary from context to context. Thus, HRM functions can be affected by organizational factors, such as company size, activity sector, type of business, relationship with other entities, general company policies and strategies, also by social, cultural, economic and political factors [17, 29, 30, 36, 60, 66, 69, 100, 103, 108, 128, 145].

## 4.5 Methodology[1]

This study is part of a broader investigation developed under an interpretative paradigm by assuming that reality is a social construction and cannot be

---

[1]This study was conducted as part of a broader project in which different themes were analysed using the same methodology. Only information corresponding to the objectives of this study was used. The broader study included the definition of research questions and the construction of the semi-structured interview guide used.

understood independently from the actors that create that reality (Urquhart, 2013). The empirical study is based on qualitative research designed to understand phenomena through the meanings that individuals attribute to them (Myers, 2011). This study used interviews for data collection purposes. Grounded Theory methodology was adopted to analyse the interviews [47, 93, 137, 155]. This analysis was confirmed and validated by a specialist in the use of Grounded Theory.

The empirical study examines 10 companies, three of which are multinational and seven of which are Portuguese. Two of the Portuguese companies are family run. Of the 10 companies, two are medium-sized companies, albeit they are leaders in their fields of business; the others are large companies (in regards to total assets, sales volume and the number of employees). The following criteria were used to select the companies: those with an HR department; those with a person responsible for HRM with the title HR manager and who effectively performed their duties; and finally, those employing an HR manager who is hierarchically and/or functionally subordinate to an administrator, general manager or superior general HR manager.

To collect the data, 257 interviews were conducted, with an average duration of 90 minutes. The interviewees represented different hierarchical levels, e.g., top managers, HR managers and their collaborators, peers of HR managers, and collaborators from different organizational functions, with or without managerial responsibilities. The interview guide consisted of 66 questions, among which were specific questions concerning the perception that the different organizational actors have about the ways of acting of HR managers. Only these questions were considered for the objective of this study. The first contact was made by phone with administrators and HR managers known to the first author, who was for a long period of time HR manager. The purpose of this first call was to explain the project and the main goals. I was also schedule a meeting for a comprehensive explanation of the research objectives, the data required, as well as for the definition of the number of interviews and the procedures to select the interviewees. The classification of the companies, their sectors of activity and country of origin, the number of interviews conducted, and the codification of interviewees are described in Table 4.1.

The interviewees covered a broad age range, from 19 to 82 years of age, and labour seniority, ranging from 3 months to 67 years, and were at different stages of their careers. The gender distribution of the interviewees was balanced across professional categories with the exception of the

**Table 4.1** Classification of the data sample

| Business Clusters | Activity | Origin | Number of Interviews by Company | Codification of the Interviewees |
|---|---|---|---|---|
| Multinational | A) Industrial | Sweden/USA | N = 27 | Sub. 1–27 |
| Companies | B) Chemistry | Germany | N = 27 | Sub. 28–54 |
| | C) Technology | Germany | N = 28 | Sub. 55–82 |
| National | D) Industrial | | N = 25 | Sub. 83–107 |
| (Portuguese) | E) Technology | | N = 25 | Sub. 108–132 |
| Companies | F) Commercial | Portugal | N = 26 | Sub. 133–158 |
| | G) Technology | | N = 25 | Sub. 159–183 |
| | H) Industrial | | N = 26 | Sub. 184–209 |
| Family | I) Commercial | | N = 24 | Sub. 210–233 |
| Companies | J) Textile | Portugal | N = 24 | Sub. 234–257 |

administrator category, for which only three of the 24 interviewees were women. Most of the interviews were carried out between 2007 and 2012, and after transcription, given the changes in the financial and political conditions of the Portuguese context, additional interviews were conducted between 2013 and 2015, confirming the previous results.

## 4.6 The Process of Devolution of HRM Responsibilities to Line Managers: Perceptions and Reactions of Different Organizational Actors about

The main objective of this study is to understand the perceptions and reactions of different organizational actors about the process of devolution of HRM responsibilities to line managers. The questions raised by this process and the perceived perceptions are very much supported by the expectations of the different organizational actors including the line managers themselves.

From the analysis of the data and from the recurrent return to them emerges characteristics, which together gave rise to categories and these have been grouped into broader domains that are presented below.

### 4.6.1 Expectations about the Nature and Degree of Intervention of Line Managers in HRM Areas

The devolution of responsibilities and functions of the manager and of the own HR management to the line managers presents itself as a process of decisions of advances and retreats, as reflected in the transcripts that follow:

> The return process seems an inevitability but it is necessary to prepare line managers for this theme and these procedures because, in general, it is where we have less knowledge and then it is what gives us more headaches. (Manager Peer)

> In my area the return process went well and today everyone agrees that it was a good decision, it took some time to fine-tune, but now we deal with some issues more effectively than the HR department treated and freed that structure to develop a whole set of value-added actions and how fashionable it is to say for more strategic actions. (Manager Peer)

> I did not feel a loss of power or influence or even a slight threat when some responsibilities were transferred to the line managers, however, the return of these responsibilities to the HR department reveals a shortage of managerial training and the process itself was not properly performed which caused some noise. (HR Manager)

This crosscutting and controversial subject in various aspects of business life in general, and of HRM in particular, will be based on two remarkable quotations from two Administrators:

> I only see benefits in returning to line managers but moderately ... and with these prepared, trained to take over and willing to take on this kind of responsibilities. (Administrator)

> ... without intending to hurt your susceptibilities, let me explain in this way what I think of the return of responsibilities to the line managers, and I am even one of those that I think is positive about it. But I say "to Caesar what belongs to Caesar, to God what is of God". Returning yes, pouring without thinking, would just screw up. (Administrator).

The return of HRM responsibilities to line managers had already occurred in all companies participating in this study, with the exception of two multi-national companies, and was taking place at the time of the empirical study in two national companies. This fact denotes that this process was made of advances and retreats, arising from the data analysis a whole set of arguments for or against, in line with the international literature.

The expectations of the top management towards the HR department (237 refs), the leadership style of the HR manager (229 refs), the qualitative and quantitative composition of the HR department (226 refs), the leadership characteristics of the various levels (216 refs), the size of the company (209 refs), its culture and values (198 refs), the technological development of systems for HRM (198 refs), and the perceived power of the HR manager and HR department (165 refs), can contribute, according to the different actors, but mainly by the holders of management positions – managers, peer managers, line managers – so that this decision to devolve HRM responsibilities to line managers will become a competitive advantage for all parties. This would contribute to the promotion of a positive perception of HRM by giving specific departments and managers in this area the opportunity to take on new and more complex responsibilities and to contribute to increasing credibility and respectability for the role played by HRM.

The analysis of the interview data allows the listing of four domains at this level: *HRM responsibilities devolution process for line managers, facilitating factors and obstacles to devolution, perception of devolution by type of actors and differentiated stages of management and HRM evolution,* and *nature of HRM.*

### 4.6.2 First Domain: Devolution Process of HRM Responsibilities for Line Managers

In this area, there are several categories: *process of the devolution, delegated responsibilities, reactions to the devolution process, perception about the devolution,* and *conditions for the devolution,* as presented in Table 4.2.

The *devolution process* presents, according to the analyzed data, two characteristics: nature and scope. Regarding the *nature* of the devolution process, two guidelines may be emphasized: first, a strategic devolution perception by the HR manager and department (56 refs); second, a tactical perception of the devolution of HRM responsibilities to line managers (49 refs).

In relation to the *scope of the process,* the data are based on a more *sporadic perceptive preponderance* (102 refs) than a more *structured* one (49

**Table 4.2**   First domain: Devolution process of HRM responsibilities for line managers

| Categories | Characteristics |
| --- | --- |
| Process of the devolution | Nature: strategic devolution or tactic devolution |
| | Scope: sporadic or structured character |
| Delegated responsibilities | Types and contents |
| Reactions to the devolution process | Types and motives of reactions |
| Perception about the devolution | Negative characterization of perceptions |
| Conditions for the devolution | Characterization of conditions as excellent or non-existent |

refs). This perceived difference may be due, on the one hand, to the advances and setbacks that this process has had and, on the other hand, to relate to some of the aspects referenced in some of the categories listed in this field, namely, *the perception about the devolution* and *the conditions for the devolution.* The manager and the HR department can have, in their strategic development plan, perfectly assumed that the HRM responsibilities devolution process can and should be carried out in a programmed way, having to prepare the line managers for this mission, as is clear from the following transcripts:

> It was easy to see, we did not have to be very smart either, just be attentive and listen to the HR manager and colleagues working in the HR department to realize that as soon as the conditions were established for us to solve some formal HRM issues, we would do. I think the strategy followed was adequate because everything was properly planned, well communicated, we had training for what is required and is also nothing transcendent. I personally like it, after all we already had to solve the problems with our employees and thus we were better prepared (Line Manger) (56 refs on a more strategic option).

> In my opinion, this process makes perfect sense and it happened here in the company, but it could have gone better if things had been better planned and organized. No one first asked my colleagues and me if we were able to get more work, then there was no care to properly explain what it was to do and how it should be done. It seems to me that this was the basics but failed and then things just did not fail because we have our professionalism and the people are our collaborators. The idea at the beginning was that the tactic was to dispatch, gilding the pill and worse saying that it were very simple things. . . you are seeing this being said by those who have maximum HRM management responsibilities, by the HR manager and my superior to be of little help ... it is to belittle one's own issues, work and ourselves. (Line Manager) (49 refs about a more tactical option)

The *sporadic or structural scope* has to do with the consolidation and maintenance of the process in the companies, which as already mentioned, occurred in most of them. However, at the time of data collection, it only existed in two companies in a consistent and lasting manner. In all companies, there are HRM tasks that line managers necessarily carry out by virtue of their position, without, however, assuming the existence of an effective devolution of HRM responsibilities. Among these tasks are procedural and control tasks, mainly in the issues of providing elements for the processing of salaries, assiduity and legal aspects.

A second category in this area has to do with the *delegated responsibilities* recognized and perceived by the different actors and based on the characteristic derived from the analysis of the data, designated by *types and contents* of delegated responsibilities. The more global and transversal perception expressed by the interviewees points to the assumption by line managers of responsibilities and/or different tasks launched by the HR department, such as: actions in the scope of *social and environmental responsibility* (208 refs), *responsibilities in the area of hygiene, health and safety at work* (159 refs), *contents of a technical and administrative nature* (153 refs), *participation in the recruitment and selection processes* (149 refs), *participation in the performance appraisal process* (82 refs), *participation in the development of internal communication systems* (76 refs), *participation in planning and delivering training* (74 refs), and *participation in situations of exercise of disciplinary power* (62 refs).

The third category, *reactions to the devolution process*, translates the history of the same process into the life of the companies, including this third category a characteristic designated by *types and reasons of the reactions* that can be classified as negative and positive. Most of the negative reactions result either from the way the process was carried out during its implementation, or from the implementation attempt with *communication deficits* (179 refs) and *lack of strategy of involvement and training of managers* (163 refs) in relation to that they were asked. The data also shows a negative reaction on the part of the line managers. This perception results from the fact that, at the time of this process development, the volume of operational and management work, as well as the structural conditions, which at the time of the return of tasks and responsibilities had not been duly assessed, as is apparent from the transcript:

> Nobody wanted to know if I was overloaded or not, you will be a pivot of HR and I am done with two operational areas and more with the pompous name of HR pivot. Of course I got upset ... now I can tell you that I do not even mind and I like, but at the beginning I did not like anything at all (Line Manager) (54 refs express this idea).

The positive reactions have to do with the way in other business realities (intra or inter-participating companies), *the devolution process was conducted*

(86 refs) and a *predisposition of the line managers for this type of functions* (38 refs). On the part of the interviewees – line managers – who currently perform functions in this area or who had already had this type of responsibility, one of the most mentioned reactions is: "... for me it was excellent, it put me even more within the questions of my people, we have close relations and this was very gratifying..." (Line Manager) (108 refs). That is, it favored the levels of social and professional interaction and increased the level of quality of interpersonal relations. Another positive reaction that emerges from the data was the *enrichment of functions and/or the widening of responsibilities* due to the centralization of some operational functions, as mentioned by the interviewees:

> The company has decided to centralize some business functions and we have more customer service responsibilities, which is no easy task. In addition, and despite a centralization of HRM, there was a need for some responsibilities to be passed on to us, and this was excellent for me because it made me feel more professionally fulfilled. (Line Manager, in a national company) (51 refs express this idea)

The professional emptying and the consequent threat of possible dismissal was compensated for "assuming a set of responsibilities referring to one of the areas in which the strategic and fundamental value for the development of the company is proclaimed internally and externally" (Line Manager).

The fourth category refers to *the perception about the devolution* by the actors belonging to the HR departments. This category relies heavily on the homogeneous perception of this particular group of actors and is broadly *negative* in general terms. There are employees in the HR departments who understand this process as a *disinclination of the job* because "line managers quickly put HRM issues second or third on their agenda." (HR Collaborator) (19 refs in 20). Also the perception of *lack of sensitivity* (18 refs in 20), *lack of aptitude* (18 refs in 20) and *lack of knowledge and willingness to receive training* (16 refs in 20) lead to negative reactions of these interviewees. Another factor that emerged from the interviews of this group was also the manifestation of "clear loss of power and undervaluation of the area" (19 refs in 20), as well as a whole set of data revealing a *feeling of threat in relation to the loss of own work* (17 refs in 20).

Finally, in *the conditions for the devolution* category, the *characterization of conditions as excellent or non-existent* emerges as a characteristic.

According to the data obtained, this category can be seen in a continuum that positions these conditions between one pole of existence of *excellent conditions for the devolution* (79 refs) to another polo where does *not exist conditions to make the devolution* (167 refs). The following quotation reflects the idea contained in most of the 167 references of the interviewees that are in the latter group:

> When someone says that we are going to assume an important responsibility at the level of a set of processes that until then were carried out by the HR department, and does not ask us if we are in a position to receive or not. We are given training during the time of a lunch, not more than an hour and want to make believe that it is not necessary because it is only paperwork ... so many questions and doubts remain. The first one I did to myself was kind: they must be playing with me, I realized they did not, that the thing was to be passed on time. If it is only paperwork, why is it not solved computerically? ... well that was too bad to be true, but that's what happened. I was hanged the monkey in my back and I watching it. This is how people are managed ... I think that this is not an example of what HRM is here in the company, it could not and should not have happened. Reducing people to paperwork should not happen if it were not for decorum, modesty, but I am the illiterate. Notice that neither my superior spoke to me and worse is that no one spoke to him and this was not treated by anyone of HR, but by a parachutist who solved everything with a *Sumol* [Orange juice brand] and a cod cake ... the process returned to where it should never have gone out in those conditions. (Line Manager, illiterate).

Other factors were referred to justify the poor conditions of implementation of this process, whether they actually occurred or were only presupposed: lack of adequate training or ministered to the people who would have to ensure the tasks/responsibilities returned (143 refs); undefined issues to be returned and degree of autonomy in its resolution (98 refs); and, no involvement of line managers in the whole process (76 refs). In the case of one of the large national companies, in which the HRM was centralized, what happened was a divestment and reduction of the area of HRM in the company units, which now has a pivot that started to accumulate HRM with the two operating areas he already had. In his words:

I am often removed from what I am doing in HRM because what I am told is that I am here to sell ... and then I have some areas of HR calling and demanding this world and the other and that if the salary processing does not close in the whole company is my responsibility. I already thought about suicide, but my guardian angel said it was not worth it (laughs) ... (Line Manager) (37 refs express this situation).

### 4.6.3 Second Domain: Facilitating Factors and Obstacles to Devolution

In this area, the data give rise to two categories: *typology of factors* and *impact of the return on results and costs*, as presented in the Table 4.3.

In terms of *typology of factors*, the analysis of interviews reveals as characteristics the *definition of the facilitating factors and obstacles* to the implementation and development of this devolution process. Regarding the *facilitating factors*, one of the factors that emerge from the analysis of the data was called *entrepreneurial empathy*. Given the right conditions for the devolution process to be properly implemented, respondents (87 refs) report that they have created a sense of understanding the role of the other and of being able to analyze the organizational reality through the perspective of the other, such as from the quotation:

The process of carrying out some of the tasks that were hitherto the responsibility of the HR department was carried out with all normality... of course there is always some fear of some mistake even because what was transferred is more of the administrative area, but that too is good because it forces us to think differently and to value the work of others. (Manager Peer).

**Table 4.3**  Second domain: Facilitating factors and obstacles to devolution

| Categories | Characteristics |
|---|---|
| Typology of factors | Definition of facilitating factors |
| | Definition of obstacles |
| Impact of devolution on results and costs | Operationalization of impacts in terms of: cost control; efficiency vs. inefficiency; relations of cooperation versus relations of competition; organizational development versus organizational entropies |

Another facilitating factor was designated by *knowledge of reality* (117 refs), translated and characterized as expressed in the following quotation:

> Those who are on the ground know their people better than anyone else and as such was unavoidable, even by the size of the company at all levels, that this process would occur. Also the number of employees of the HR department, and especially the investments made by the company in the computer system, it is normal that the company intends to monetize its resources and hence some tasks are transferred to the line managers and to us [managers' peers]. The company has this culture of knowledge multiplying and as it gives proper training things become easier to accept. Of course it does not always go well and we have been in this process for half a dozen years, but the final balance is positive... Before my line managers were not able to inform me immediately about who was missing work, now they push a button and tell me immediately and I can also access. Another example, a person may miss two days to work but there may be another person who systematically arrives late and at the end of the month ends up working less than the one that was missing, this is now visible and useful. So, I think that the devolution should be understood as an improvement in our management and also of line managers and not a simple act of processing, as the examples I gave you, because that would be the way HR would handle matters for lack of time from the HR department. (Manager Peer).[2]

Still in the same category, the data allow to list another facilitating factor called *perception of interdependence of functions* (89 refs). This factor refers to the mentioned situation in which are still the activities of more technical and administrative character that are object of greater devolution. The responsibilities of the HR manager are not very transferable, except that the HR manager and the HR department, in order to carry out their duties, inevitably need the collaboration of third parties, recognizing the same need, as follows:

> It should be stressed that the manager and the HR department, when they fail to fulfil certain responsibilities, do not mean that

---

[2]These ideas were expressed by a large majority of the organizational actors with leadership responsibilities, with 117 references, including some technical staff.

they do not have a right to exist, but that this need for existence is reinforced. On the other hand, what is transferred to us are administrative aspects and it was bad if HRM were just paperwork, which would be the easiest... Now the HR manager maintains the attitude of cooperation and partnership that he maintained with us, for example in the selection of people, at least at a certain level we have to collaborate the two, in the performance evaluation must give instructions and get everyone to get to reduce to the maximum possible subjectivity and the hypotheses of injustice... Teamwork, it has to be. (Manager Peer)

There are factors that are referred to as facilitators or as obstacles depending on the type of conditions created when the HRM responsibilities devolution process is implemented, as presented in Table 4.4.

In the category *impact of the return on results and costs,* the characteristic that arises was denominated by *operationalization of the impacts.* The analysis of the data reveals the existence of differentiated perceptions. One of the most referenced aspects has to do with the type of business of the company and the organizational culture, in the sense that this devolution can provide *control of costs* in terms of rationalization and reduction thereof.[3] Three other aspects of this category can be operationalized in three continuums: efficiency versus inefficiency (74 refs versus 32 refs); relations of cooperation versus relations of competition (156 refs versus 42 refs); and organizational development versus organizational entropies (179 refs versus 67 refs).

**Table 4.4**   Types of factors: Facilitators/obstacles

| Types of Factors | Facilitators | Obstacles |
|---|---|---|
| Type of line managers and their degree of preparation | 107 | 92 |
| Availability of line managers to take on new responsibilities. | 69 | 145 |
| Complexity of roles and responsibilities to be returned | 76 | 128 |
| Leadership style and quality and number of HR | 103 | 54 |
| Geographical dispersion of the company | 167 | 76 |
| Technological development in HRM | 133 | 32 |
| Formalization of processes, techniques, practices and procedures | 62 | 78 |

Note: References are not mutually exclusive, meaning that the same participant may have referred to the same factor as facilitator and as an obstacle.

---

[3] 138 references go in the direction of achieving this control and decrease of costs and 102 references refer the opposite effect, in these are included 19 references in 20 possible of employees of the department of HR.

Regarding this issue, it should also be noted that this process is seen by the interviewees as not questioning the role of the manager and HR department, and is perceived as an opportunity to develop this area and its protagonist (217 refs). There is also a perceived threat (36 refs), particularly from HR department employees and some line managers directly affected by the process, who have or have had in the past to develop this function in accumulation with the management of other operational areas. These line managers fear that "if things go less well or badly I'm the one who will suffer the consequences ... things have fallen from parachutes, they gilded me the pill, but it may well be a poisoned gift." (Line Manager). A final perception taken in a transversal way, and provided that this whole process is properly conducted from the top to the bottom, is the feeling of inevitability (158 refs) for the manager and the HR department as a privileged form of development of this area of management in the company, its credibility and valuation at the most diverse levels.

### 4.6.4  Third Domain: Perception of Devolution by Type of Actors and Differentiated Stages of Management and HRM Evolution

In this area, the emerging categories have to do with the type of actors, namely in terms of *professional categories*. Data analysis reveals that this is the differentiating variable particularly in the following organizational actors: administrators, undifferentiated collaborators, line managers and HR managers. The *degree of evolution or maturity of management* in the company in general and the *degree of evolution or maturity of HRM*, in particular, to frame this process, as presented in Table 4.5, also appear as mediating factors.

The analysis of the interviews reflects that the issue of devolution, although complex, is more peaceful at the level of the functional ends of the company, in terms of organizational actors. Thus, administrators see the possibility of increasing the level of intervention of the line managers with their work teams, the possible reduction of structural costs, and the release of

**Table 4.5**  Third domain: Perception of devolution by type of actors and differentiated stages of management and HRM evolution

| Categories | Characteristics |
| --- | --- |
| Professional categories | Degree of evolution or maturity of management in the company |
| | Degree of evolution or maturity of HRM in the company |

the HR manager and the respective department to different responsibilities. For example:

> This is not a simple process and sometimes it has to be handled with great care as it is a process of management of multiple sensitivities, but in our view, it is an inevitability and will increase credibility and recognition of HRM and its professionals. (Administrator)

In turn, the undifferentiated operational and administrative collaborators see this process of devolution to line managers as a way to guarantee a greater effectiveness of the application of HRM practices, techniques and programs, given the greater proximity of the line managers with the work teams. The problem occasionally raised in the interviews with many participants (148 refs) occurs, on the one hand, given the lack of training and aptness of the line managers for this area and, on the other hand, not being recognized by their superiors this work in terms of performance appraisal. This leads many managers to reduce both their degree of performance and intervention as well as the priority they assign to their assigned tasks and responsibilities. Exemplifying:

> I agree with the assignment of HRM tasks, because we are the ones who are closer to the people and already in day-to-day we solve many of the issues, of course in articulation with the HR manager and HR department. (Line Manager)

> Totally agree, as long as we have training otherwise, we just get enough rope to hang ourselves. (Line Manager).

These arguments are exactly those used by some HR directors, as follows from the following quotation:

> ... in some countries the issue of the return was abandoned, at a time when many Portuguese companies want to do it for the sake of reducing costs ... reduction ... but it is like many measures, save in pennies and spend in the millions. The experiences that occur translate into more costs, demotivation and lack of sensitivity. (HR manager of a Portuguese company with two-headed leadership in HR).

For line managers this devolution is variable in terms of the degree of acceptance. On the one hand, they consider the devolution of HRM responsibility very dependent on the state of development of HRM in the company. On the other hand, it is dependent on the level of centralization of various functions that may have occurred or is expected to occur (not only HRM functions, but also administrative functions, accounting, logistics, etc.). For example, in the case of the two companies that centralized the HRM, centralization of other management areas was also due to the reorganization of operational processes or the introduction of new technologies, leading to a functional emptying and causing the operational units to be reduced to two activities: commercial and management of work teams. This has meant that the appropriation of HRM responsibilities, through the devolution that the centralized HR management implied, was perceived by the line managers as something that "fills them better and completes their job" (Line Manager), since they consider that "It is an inevitability this devolution in which we all win" (Line Manager). In other cases, line managers may experience adverse reactions because they feel unprepared for these requirements and continue to have a highly charged role with operational and management tasks and responsibilities.

Another factor that generates controversy between managers and employees of HR departments and line managers has to do with the level and areas of operation. Line managers sometimes do not want to take on the difficult and embarrassing situations of HRM, which leads some HR managers to refer to the unusual selectivity of line managers, stating that "... to deal with pleasant matters there is no shortage of those who want, now when issues are problematic, they soon send to HR" (HR Manager). For their part, line managers in these most problematic situations tend to assume positions similar to those referenced in the following citations:

> This type of process – admonition of workers – is not part of the core of our function. (Line Manager)

> I believe that these complicated situations have to be dealt with by the HR department and the HR manager for two reasons: firstly, they are trained to do this and, secondly, because after the situation is solved we have to continue working with people, which would lead to confusing situations. (Line Manager)

Hierarchically superior leaders (e.g., manager peers) have an ambivalent feeling. On the one hand, the accomplishment of tasks and the assumption of responsibilities of HRM bring a greater power of management. On the other hand, given that the results in this area of management are not usually immediate, there is a tendency to devalue and depreciate these responsibilities. In relation to the priority given by managers to the devolution process for the line managers under their management, it varies between two poles: high priority (104 refs) and low priority (89 refs). The high priority given by the managers occurs when they perceive the good assumption of the return process by the line managers under their management, seeing in this fact a hypothesis of capitalizing advantages with the HR department, namely in terms of evolution and career management. The low priority occurs when there is a lack of immediate results of HRM activities along with eventual criticism from the top of the organization to operational performance. As explained, given the "... deviation of concentration of the line managers in the operational areas leads to the reduction of results, it is necessary focus on the tasks in which we are actually evaluated" (Manager Peer) (89 refs). Underlying this category and the two interdependent characteristics there is a perceptive and instrumental positioning of this type of organizational actors in relation to the devolution process.

### 4.6.5 Fourth Domain: Nature of HRM

In this domain the category that emerges from the data has to do with HRM and the *typology of the function*, which can be a *function of specialists* (157 refs) or a *function of generalists* (93 refs). This double typology arises mediated by three characteristics: the *size of the company*, the *organizational structure* and the *implementation of certain projects*, as presented in Table 4.6.

The citation that is presented reflects the process of evolution of the size of the company and its own organizational structure:

> We felt the need to create the HR department from the moment we reached the 120 employees. So far, for the sake of temperament and taste, I was the one who had the issues of HRM, but from a given moment it was no longer feasible. We had a need and we looked for someone senior who dominated the spectrum of the areas of

**Table 4.6** Fourth domain: Nature of HRM

| Categories | Characteristics |
|---|---|
| Typology of the function | Size of the company |
| | Organizational structure |
| | Implementation of certain projects |

HRM because we could not and cannot have a heavy HR structure, the company cannot have determined types of costs. So I would say that our HR manager is generalist but very good in all matters, she already has a seniority. Then we have one person for administrative issues and another who is a psychologist responsible for recruitment and selection and part of the training. (Administrator)

On the other hand, there are still moments in the life of companies that coincide with the implementation of certain structural projects, such as the quality certification process. These structural processes, on the one hand, condition the way in which HRM is perceived and, on the other hand, have an impact on the issue of the devolution of operational responsibilities to line managers, and imply the assumption of an HRM mentality by the different types of actors, as presented:

There are moments in the life of the company where there is a need to have an even minimal HR policy structure, as happened in the quality certification process... (Manager Peer).

...the HR structure emerged to meet the needs that had been placed in the scope of the quality certification process, namely in terms of descriptions, analysis and qualification of functions, description of some procedures and development of training processes. Subsequently, the need to have and maintain an HR structure was perceived, since, on the one hand, certain requirements were created and, on the other hand, new needs were created and situations were resolved that made the HR structure a permanent necessity. Initially on a generalist basis due to limitations of financial resources and more recently, without losing this generalist base, it was possible to endow it with some specialization. (Administrator)

The state of evolution of the HRM itself may lead one to think that at first this is a function of specialists, as a way of promoting a greater and more consistent maturity of this area of management. However, there is also the view that HRM being so comprehensive and crosscutting cannot be exclusively in the "hand" of a single person or a single department, but rather involve the entire organization. This prospect of return does not make GRH a generalist area, but it should help to emphasize the need for and imperative commitment and accountability of the entire organization in this regard.

Finally, the "Taylorization" of HRM in specific aspects of this area, such as administrative and legal issues, through the implementation of very strict procedures, total computerization and the non-existence of a critical reflection of data from a more qualitative point of view, can diminish this area of management. According to the interviewees, this situation is "something that can be very reductive because there can be a risk of losing sensitivity to deepen these issues, which may in turn prove to be excellent symptoms and barometers of the social climate of an organization". (Manager Peer)[4]

## 4.7 Conclusion

The issue of devolution of HRM roles and responsibilities to managers in other areas of management needs to be adequately contextualized in terms of variables such as: degree of development of HRM in an organization; business strategy and evolutionary trends; the perception that top management has about this process; degree of preparation and qualification of the different organizational actors; communicational maturity of the organization; competence in the management of change processes, among many other contextual dimensions.

In the same organization it is possible to find arguments in favor and against the devolution, and even change of opinions and behaviors according to personal strategies and tactics. The study highlights that the same or similar arguments can be used to justify both the devolution and maintenance of HRM responsibilities. However, it is important to consider that strategic decisions to centralize and decentralize the different areas of management by top management can be a decisive factor in the devolution of HRM tasks, functions and responsibilities to line managers.

Another aspect that emerges from the data has to do with the very question of status of the profession and even the maintenance of employment by the persons involved in HR management. Many of HRM's administrative and technical collaborators do not always express this fear directly, but they list a whole set of arguments against the devolution decision. This argument against devolution is based on the disbelief that the HR function may suffer; in the lack of training and even attention and vocation of the line managers for these issues; in the increase of the number of errors and consequent increase of costs and of loss of confidence. They also point out that line managers "just want to get the good part of the job" and want to "discard the less friendly parts" and that "the hateful part of the decisions is always for us" (HR collaborator), thus justifying its position against the devolution.

---

[4] 148 references express this idea transversal to all types of employees and in all companies.

It is also apparent from the data that the participating HR managers have a more favorable position regarding the devolution than the HRM technical and administrative staff. This attitude may result from a stronger sense of professional identity on the part of the technical and administrative employees who feel a more concrete threat to their functions, emptying of position and consequent loss of employment. On the other hand, the line managers who are favorable to receive HRM responsibilities point to the issue of proximity and knowledge of their people and teams as justifications for such process. However, for HR technical and administrative collaborators, the centralization of many responsibilities that line managers had in their functional content for core areas means enrichment of their job and thus reduces the fear expressed by those employees of emptying the job and removes the threat of unemployment.

An aspect that also arises from the data has to do with the perspective of some of the hierarchical superiors of the line managers, who expressed perceptions corresponding to contradictory ways of analyzing this devolution process, in agreement with Sikora and Ferris (2014). On the one hand, they point out the inevitability of this process to happen. On the other hand, they refer and make it a point to remind line managers that the focus of their activity is not the resolution of HRM issues, since they will be evaluated by the performance and results of their area of operational intervention. This contradiction leads some line managers to express a difficulty in knowing how to deal with this process of devolution and thus assume defensive positions seeking to safeguard themselves from a *crossfire*.

The argument of the inevitability of the return process is transverse, but particularly emphasized by the top managers and also by the peer managers of the HR manager. The justification for this inevitability arises as a guarantee and a prerequisite for better efficiency and greater effectiveness in dealing with issues due to the proximity of line managers with people and a better knowledge of their issues and problems. In turn, the manager and the HR department are freer and more available for other issues considered as more important and have a more relevant strategic role in the organization.

In addition, HRM is characterized as a crosscutting area with relevance to organizations for the role they can play in terms of empowering people with the skills and abilities and qualities that enable them to play a role in the organization's sustained growth and development, and making decisions in line with this sustainability. A company will be all the better the better the people that integrate it and it is in this dimension that the HRM with its specifics can take on a differentiating nature and add value in terms of organizational performance, both at strategic and operational level. However, this study shows that some HR managers are seen and perceived by others as given credibility to their function not because of its specificities, but through a discourse that is validated in function of the dominant discourse and the results.

However, there is still a reality of line managers composed of people who, being excellent technical and administrative collaborators, ascended to management positions without the skills and qualities needed to coordinate people and teams. In short, in these cases, companies can develop rather perverse processes in which, with losses to themselves, they have a weak manager and lose a good technical or administrative professional. Therefore, as practical implications of the study, as highlighted by Kinicki et al. [90]; see also [156], several performance management behaviors could be integrated in (line) managers' training program to guarantee the success of the devolution process, such as: goal-setting, communication, feedback delivery, coaching, providing consequences and the monitoring of performance expectations.

Managers also suffer from a problem in this devolution process, which has to do with a high desire to control the entire activity of the people they supervise. This may lead line managers to refrain from taking on the responsibilities of the devolution process, to disengage themselves and to disassociate themselves from the commitment that is essential in this process.

Well-structured communication systems; leadership and change management skills; creation of organizational empathy; prepare this process properly and involve people by giving them relevant training, are fundamental conditions for the success of the devolution process. As argued by Sikora and Ferris (2014), it is important that organizations and their leaders acknowledge that although the development of HR practices is the responsibility of the HR department, the implementation of those practices is the responsibility of line managers. In fact, line managers are seen as the key actors in HRM, since they hold a vital position in the implementation of HRM practices [9, 156]. Accordingly, "line managers need to be competent and motivated to effectively carry out their HRM responsibilities. In addition, line managers need sufficient support from HR professionals to provide them with the necessary HR skills and the proper encouragement to perform their HR role" [9].

An organization is all the better the better its people are and the more someone wants to progress in professional terms, the more they will have to surround themselves with people who add value to their knowledge, their doing, their being and their evolution. This idea is based on the assumption of conscious, competent and responsible involvement of managers from different areas with HRM responsibilities. It takes a reciprocal and empathic understanding of the role of each person in the organization to build a more motivating and simultaneously more sustainable organization!

We leave in a world of perceptions and future studies could contribute to a better understanding of the perceived line managers' motivations to implement HRM practices, and how they are influenced by personal characteristics, which could help in selection procedure of line managers [120]. Also, it is important to study the perceptions by different actors on how well line managers are actually performing their HRM responsibilities [9].

## Acknowledgements

Delfina Gomes acknowledges that this study was conducted at the Research Center in Political Science (UID/CPO/0758/2019), University of Minho/University of Évora, and was supported by the Portuguese Foundation for Science and Technology and the Portuguese Ministry of Education and Science through national funds.

## References

[1] Aguinis, H. (2013). *Performance management*. Upper Saddle River, NJ: Pearson Prentice Hall.

[2] Aguinis, H., Joo, H., & Gottfredson, R. (2012). Performance management universals: Think globally and act locally. *Business Horizons, 55*, 385–392.

[3] Almeida, J.A. (2012). Os profissionais de gestão de recursos humanos: competências e espaços de reconhecimento profissional. *In* T. Carvalho, R. Santiago & T. Caria (Eds), *Grupos Profissionais, Profissionalismo e Sociedade do Conhecimento: Tendências, Problemas e Perspectivas* (pp. 97–108). Porto: Edições Afrontamento.

[4] Armstrong, M. (2000). The name has changed but has the game remained the same?. *Employee Relations*, 22(6), 576–593.

[5] Armstrong, M., & Taylor, S. (2017). *Armstrong's handbook of human resource management practice*. London: Chartered Institute of Personnel and Development.

[6] Attali, J. (1999). *Dicionário do Século XXI*. Lisboa: Multitipo Artes Gráficas.

[7] Beaumont, P.B. (1995). The U.S. human management literature: a review. *In* G. Salaman (Ed), *Human Resource Strategies* (pp. 20–37). London: Sage.

[8] Becker, M.C. (2004). Organizational routines: a review of the literature. *Industrial and Corporate Change*, 13(4), 643–678.

[9] Beeck, S., Winen, J., & Hondeghem, A. (2017). Effective HRM Implementation by Line Managers: Relying on Various Sources of Support. International Journal of Public Administration, 40(2), 192–204.

[10] Beer, M. (1997). The transformation of human resource function: resolving the tension between a traditional administrative and a new strategic role. *Human Resource Management*, 36(1) 49–56.

[11] Benjamin, J. & Louw-Potgieter, J. (2008). Professional work and actual work: the case of industrial psychologists in South Africa. *South African Journal of Psychology*, 38(1): 116–135.

[12] Bloom, N., & Van Reenen, J. (2011). Human resource management and productivity. *In* O. Ashenfelter & D. Card (Eds), *Handbook of Labor Economics* (pp. 1697–1769) (Vol. 4, Part B). North Holland: Elsevier.

[13] Bloom, N., Genakos, C., Sadun, R., & Van Reenen, J. (2012). Management practices across firms and countries. *Academy of Management Perspectives*, 26(1), 12–33.

[14] Bonache, J. (2007). Los recursos humanos en la internacionalización del Grupo Santander: objetivos, logros e retos. *Universia Business Review*. Número especial 150 aniversário Banco Santander primer trimester (1° trimeste 2008), 46–59.

[15] Bondarouk, T., Looise, J.K., & Lempsink, B. (2009). Framing the implementation of HRM innovation. HR professionals vs line managers in a construction company. *Personnel Review*, 38(5), 472–491.

[16] Boon, C., Den Hartog, D. N., Boselie, P., & Paauwe, J. (2011). The relationship between perceptions of HR practices and employee outcomes: examining the role of person-organization and person-job fit. *International Journal of Human Resource Management*, 22(1), 138–162.

[17] Bournois, F. (1991). Pratiques de gestion des RH en Europe: données comparées. *Revue Française de Gestion,* March-May, 68–83.

[18] Boxall, P. (1992). Strategic human resource management: beginnings of a new theoretical sophistication? *Human Resource Management Journal*, 2(3): 60–79.

[19] Boxall, P., & Purcell, J. (2000). Strategic human resource management: where have we come from and where should we be going? *International Journal of Management Reviews* 2(2), 183–203.

[20] Brandão, M.A., & Parente, C. (1998). Configurações da função pessoal: as especificidades do caso português. *Organizações e Trabalho*, 20, 23–40.

[21] Bratton, J., & Gold, J. (1999). *Human Resource Management: Theory and Practice*. London: Macmillan Business.

[22] Breitfelder, M., & Dowling, D. (2008). Why did we ever go into HR? *Harvard Business Review*, 86(7/8), 39–43.

[23] Brewster, C. (1997). A profissão de gerir pessoas e talentos. *18° Congresso Europeu de Recursos Humanos* (pp. 7–13). Lisboa: APG.

[24] Brewster, C. (1999). Strategic human resource management: the value of different paradigms. *Management International Review*, 39(3), 45–64.

[25] Brewster, C., & Hegewisch, A. (1993). Human resource management in Europe: issues and opportunities. *In* C. Brewster & A. Hegewisch (Eds), *Policy and Practice in European Human Resource Management: The Price Waterhouse Cranfield Survey* (pp. 1–21). London: Routledge.

[26] Brewster, C., & Hegewisch, A. (1994). *Policy and Practice in European Human Resource Management: The Price Waterhouse Cranfield Survey*. London: Routledge.

[27] Buhler, P.M. (2008). Managing in the new millennium. The skills gap: how organizations can respond effectively. *Super Vision*, 69(1), 19–22.

[28] Buyens, D., & De Vos, A. (2001). Perceptions of the value of the HR function. *Human Resource Management*, 11(3), 70–89.

[29] Cabral-Cardoso, C. (1999). Gestão de recursos humanos: evolução do conceito, perspectivas e novos desafios. *In* M.P. Cunha (Ed), *Teoria organizacional: Perspectivas e Prospectivas* (pp. 225–249). Lisboa: D. Quixote.

[30] Cabral-Cardoso, C. (2004). The evolving Portuguese model of HRM. *International Journal of Human Resource Management*, 15(6), 959–977.

[31] Cadin, L., Guérin, F., Pigeyre, F., & Pralong, J. (2012). *Pratiques et Éléments de Théorie: Gestion des Ressources Humaines* (4ª ed.). Paris: Dunod.

[32] Caldwell, R. (2002). A change of name or a change of identity? Do job titles influence people management professionals' perceptions of their role in managing change? *Personnel Review*, 31(5/6), 693–709.

[33] Caldwell, R. (2003). The changing roles of personnel managers: old ambiguities, new uncertainties. *Journal of Management Studies*, 40(4), 983–1004.

[34] Câmara, P.B., Guerra, P.B. & Rodrigues, J.B. (1998). *Humanator: Recursos Humanos e Sucesso Empresarial*. Lisboa: D. Quixote.

[35] Canavarro, J.M. (2000). *Teorias e Paradigmas Organizacionais*. Coimbra: Quarteto Editora.

[36] Cardon, M., & Stevens, C. (2004). Managing human resource in small organizations: what do we know? *International Journal of Human Resource Management*, 14(3), 295–324.

[37] Carr-Ruffino, N. (2006). *Managing Diversity: People Skills for a Multicultural Workplace* (7*th* Ed.). Boston: Pearson.

[38] Carvalho, T., Santiago, R., & Caria, T. (2012). *Grupos profissionais, Profissionalismo e Sociedade do Conhecimento: Tendências, Problemas e Perspectivas*. Porto: Edições Afrontamento.

[39] Cascio, W.F. (1998). *Managing Human Resources – Productivity, Quality of Work Life, Profits* (5*th* Ed.). New York: McGraw-Hill.

[40] Cascio, W.F. (2000). *Costing Human Resources – The Financial Impact of Behavior in Organizations* (4a Ed.). South Western: College Publishing.

[41] Chiavenato, I. (1987). *Teoria Geral da Administração*. São Paulo: McGraw-Hill.

[42] Chiavenato, I. (1999). *Gestão das Pessoas: O Novo Papel dos Recursos Humanos nas Organizações*. Rio de Janeiro: Campus.

[43] Clark, J. (1993). Personnel management, human resource management and technical change. *In* J. Clark (Ed*). Human Resource Management & Technical Change* (pp. 1–20). London: Sage Publications.

[44] Colbert, A., Rynes, S., & Brown, K. (2005). Who believes us? Understanding managers' agreement with human resource research findings. *The Journal of Applied Behavioral Science*, 41(3), 304–325.

[45] Conner, J., & Ulrich, D. (1996). Human resource roles: creating value, not rhetoric. *Human Resource Planning*, 19(3), 38–49.

[46] Cooper, D., & Thatcher, S. M. B. (2010). Identification in organizations: the role of self-concept orientations and identification motives. *Academy of Management Review*, 35(4), 516–538.

[47] Corbin, J., & Strauss, A. (2008). *Basics of Qualitative Research: Techniques and Procedures for Developing Grounded Theory* (3rd Ed.). Throusand Oaks, CA: Sage.

[48] Correia, M.F., Cunha, R.C., & Scholten, M. (2013). Impact of M&as on organizational performance: the moderating role of HRM centrality. *European Management Journal*, 31(4), 323–332.

[49] Cotrim, T. (2001). O que mudou nos recursos humanos? *Exame*, 188, 82–84.

[50] Cunha, M.P., & Rego, A. (2005). *Liderar*. Lisboa: Dom Quixote.

[51] Cunha, M.P., & Rego, A. (2010). Complexity, simplicity, simplexity. *European Management Journal*, 28(2), 85–94.

[52] Cunha, M.P., Rego, A., Cunha, R.C., & Cabral-Cardoso, C. (2003). *Manual de Comportamento Organizacional e Gestão*. Lisboa: RH Editora.

[53] Cunha, P.C., Rego, A., & Cunha, R.C. (2007). *Organizações Positivas*. Lisboa: D. Quixote.

[54] Currie, G. (2006). Reluctant but resourceful middle managers: the case of nurses in the NHS. *Journal of Nursing Management*, 14(1), 5–12.

[55] Davila, J., Foster, G., & Jia, N., (2010). Building sustainable high-growth startup companies: management systems as accelerator. *California Management Review*, 52(3), 79–105.

[56] DeNisi, A., & Murphy, K. (2017). Performance appraisal and performance management: 100 years of progress? *Journal of Applied Psychology, 102*, 421–433.

[57] DeNisi, A., & Smith, C. (2014). Performance appraisal, performance management, and firm level performance: A review, a proposed model, and new directions for future research. *The Academy of Management Annals, 8*, 127–179.

[58] Des-Horts, C.H.B. (1987). Typologie des pratiques de gestion des ressources humaines. *Revue Française de Gestion*, 65/66, 149–155.

[59] Des-Horts, C.H.B. (1988). *Vers une Gestion Stratégique des Ressources Humaines*. Paris: Les Editions d'Organization.

[60] Dewettinck, K., & Remue, J. (2011). Contextualizing HRM in comparative research: the role of the cranet network. *Human Resource Management Review*, 21(1), 37–49.

[61] Dolan, S.I., Eisler, R., & Raich, M. (2010). Managing people and human resources in the XXI Century. Shifting paradigms, emerging roles, threats and opportunities. *Effective Executive*, 13(6), 39–46.

[62] Drory, A., & Vigoda-Gadot, E. (2010). Organizational politics and human resource management: a typology and the Israeli experience. *Human Resource Management Review*, 20(3), 194–202.

[63] Edgley-Pyshorn, C., & Huisman, J. (2011). The role of the HR department in organizational change in a British university. *Journal of Organizational Change Management*, 24(5), 610–625.

[64] Evetts, J. (2006). Short Note: The sociology of professional Groups, New Directions. *Current Sociology*, 54(1), 133–143.

[65] Evetts, J. (2012). Sociological analysis of the new professionalism: knowledge and expertise in organizations. In T. Carvalho, R. Santiago & T. Caria (Eds), *Grupos Profissionais, Profissionalismo e Sociedade do Conhecimento: Tendências, Problemas e Perspectivas* (pp. 13–27). Porto: Edições Afrontamento.

[66] Farndale, E. (2005). HR department professionalism: a comparison between the UK and other European countries. *International Journal of Human Resource Management*, 16(5), 660–675.

[67] Farndale, E., & Brewster, C. (2005). In search of legitimacy: personnel management associations worldwide. *Human Resource Management Journal*, 15(3), 33–48.

[68] Farndale, E., Paauwe, J., & Boselie, P. (2010). An exploratory study of governance in the intra-firm human resources supply chain. *Human Resource Management*, 49(5), 849–868.

[69] Farndale, E., Paauwe, J., Morris, S.S., Stahl, G., Stiles, P., Trevor, J., & Wright, P. (2010). Context-bound configurations of corporate HR functions in multinational corporations. *Human Resource Management*, 49(1), 45–66.

[70] Fombrun, C.J., Tichy, N.M., & Devanna, M.A. (1984*). Strategic Human Resource Management.* New York: Jonh Wiley & Sons.

[71] Francis, F. & Keegan, A. (2006). The changing face of HRM: in search of balance. *Human Resource Management Journal*, 16(3), 231–249.

[72] Fuller, J.B., Hester, K., Barnett, T., Frey, L., Reylea, C., & Beu, D. (2006). Perceived external prestige and internal respect: new insights into organizational identification process. *Human Relations*, 59(6), 815–846.

[73] Gilbert, C., De Wine, S., & Sels, L. (2011a). Antecedents of front-line managers' perceptions of HR role stressors. *Personnel Review*, 40(5), 549–569.

[74] Gilbert, C., De Wine, S., & Sels, L. (2011b). The influence of line managers and HR department on employees' affective commitment. *International Journal of Human Resource Management*, 22(8), 1618–1637.

[75] Gilbert, P. (1999). La gestion prévisionnelle des ressources humaines: histoire et perspectives. *Revue Française de Gestion*, 124, 66–75.

[76] Gilmore, S., & Williams, S. (2007). Conceptualising the "personnel professional": a critical analysis of the chartered institute of personnel development's professional qualification scheme. *Personnel Review*, 36(3), 398–414.

[77] Griswold, W. (2008). *Cultures and Societies in a Changing World* (3a Ed). Thousand Oaks, CA: Pine Forge Press.

[78] Guest, D. (1997). Human resource management and performance. *International Journal of Human Resource Management*, 8(3), 263–276.

[79] Guest, D. (1999). Human resource management and industrial relations. *In* M. Poole (Ed), *Human Resource Management: Critical Perspectives on Business and Management* (pp. 94–113). London: Routledge.

[80] Guest, D. (2001). Human resource management: when research confronts theory. *International Journal of Human Resource Management*, 12(7), 1092–1106.

[81] Guest, D. (2010). Human resource management and performance: still searching for some answers. *Human Resource Management Journal*, 21(1), 3–13.

[82] Hall, L., & Torrington, D. (1998). Letting go or holding on - The devolution of operational personnel activities. *Human Resource Management Journal*, 8(1), 41–55.

[83] Harris, L., Doughty, D., & Kirk, S. (2002). The devolution of HR responsibilities - Perspectives from the UK's public sector. *Journal of European Industrial Training*, 26(5), 218–229.

[84] Hendry, C. (1990). The corporate management of human resources under conditions of decentralization. *British Journal of Management*, 1(2), 91–103.

[85] Holt-Larsen, H., & Brewster, C. (2003). Line management responsibility for HRM: what's happening in Europe? *Employee Relations*, 25(3), 228–244.

[86] Hutchinson, S., & Purcell, J. (2010). Managing ward managers for roles in HRM in the NHS: overworked and under-resourced. *Human Resource Management Journal*, 20(4), 357–374.

[87] Jackson, S., Schuler, R., & Werner, S. (2009). *Managing Human Resources*. South-Western: Cengage Learning.

[88] Kahneman, D. (2012). *Pensar Depressa e Devagar*. Lisboa: Círculo dos Leitores

[89] Khan, A., & Khan, R. (2011). The dual responsibility of the HR specialist. *Human Resource Management International Digest*, 19(6), 37–38.

[90] Kinicki, A., Jacobson, K., Peterson, S., & Prussia, G. (2013). Development and validation of the performance management behavior questionnaire. *Personnel Psychology*, 66, 1–45.

[91] LaMarsh, J. (2004). Building a strategic partnership and HR's role of change manager. *Employment Relations Today*, 31(3), 17–27.

[92] Langley, A., & Tsoukas, H. (2010). Introducing "perspectives on process organization studies". *In* T. Hernes & S. Maitlis (Eds), *Process, Sensemaking & Organizing* (pp. 1–26). Oxford: Oxford University Press.

[93] Laperrière, A. (2010). A teorização enraizada (grounded theory): procedimento analítico e comparação com outras abordagens similares. *In* J. Poupart, J.-P. Deslauriers, L. Groulx, A. Laperrière, R. Mayer & A. Pires (Eds), *A pesquisa qualitativa: Enfoques epistemológicos e metodológicos* (pp. 353–385), Tradução de Ana Cristina Nasser (2a Ed.). Petrópolis, RJ: Vozes.

[94] Legge, K. (1989). Human resource management: a critical analysis. *In* J. Storey (Ed.), *New Perspectives on Human Resource Management* (pp. 19–40). London: Routledge.

[95] Legge, K. (1995). *Human Resources Management: Rhetorics and Realities*. London: The Macmillan Press.

[96] Legge, K. (1999). Representing people at work. *Organization*, 6(2), 247–264.

[97] Lemelin, M., & Rondeau, A. (1990). Les nouvelles stratégies de gestion des ressources humaines. *In* M. Leclerc (Ed), *Les Nouvelles Stratégies de Gestion des Ressources Humaines*. Québec: Presses de L' Université du Québec.

[98] Lemmergaard, J. (2009). From administrative expert to strategic partner. *Employee Relations*, 31(2), 182–195.

[99] Lengnick-Hall, M.L., & Aguinis, H. (2012). What is the value of human resource certification? A multi-level framework for research. *Human Resource Management Review*, 22(4), 246–257.

[100] Ling, Y., Simsek, Z., Lubatkin, M.H., & Veiga, J.F. (2008). The impact of transformational CEOs on the performance of small to

medium–sized firms: does organizational context matter? *Journal of Applied Psychology*, 93(4), 923–934.

[101] Martins, M. (2004). Perspectivas na acção social. *In* Câmara Municipal da Covilhã (Ed.), *Século XXI Perspectivas* (pp. 89–107). Lisboa: Editorial Presença.

[102] Maxwell, G., & Watson, S. (2006). Perspectives on line managers in human resource management: Hilton International's UK Hostels. *International Journal of Human Resource Management*, 17(6), 1152–1170.

[103] McGraw, P. (2004). Influences on HRM practices in MNCs: a qualitative study in the Australian context. *International Journal of Manpower*, 25(6), 535–546.

[104] McShane, S.L. (1995). *Canadian Organizational Behaviour*. Boston: Irwin.

[105] Miles, R., & Snow, C. (1978). *Organizational Strategy, Structure and Process*. New York: Mcgraw–Hill.

[106] Miles, R., & Snow, C. (1984). Designing strategy human resources systems. *Organizational Dynamics*, 13(1): 36–52.

[107] Mintzberg, H. (2006). *Le Manager au Quotidien, Les 10 Rôles du Cadre*. Paris: Ed. d'Organisation.

[108] Moura, E. (2000). *Gestão dos Recursos Humanos: Influências e Determinantes do Desempenho*. Lisboa: Edições Sílabo.

[109] Myers, M.D. (2011). *Qualitative Research in Business and Management*. London: Sage.

[110] Nehles, A.C., Van Riemsdijk, M.J., & Looise, J.K. (2006). Implementing human resource management successfully: a first-line management challenge. *Management Review*, 17(3), 256–273.

[111] Ng, L.C. (2011). Best management practices. *Journal of Management Development*, 30(1), 93–105.

[112] Oliveira, M.J. (2013). Novas vagas e outros destinos. *In* A. Barreto (Ed.), *Adeus Liberdade. Viva a Liberdade* (pp. 88–95). Lisboa: Fundação Francisco Manuel dos Santos.

[113] Paauwe, J., & Boselie, P. (2003). Challenging strategic human resources management and the relevance of the institutional setting. *Human Resource Management Journal*, 13(3), 56–70.

[114] Parry, E., Stavrou-Costea, E., & Morley, M.J. (2011). The Cranet International Research Network on Human Resource Management in retrospect and prospect. *Human Resource Management Review*, 21(1),1–4.

[115] Patti, A., Fok, L., & Hartman, S. (2004). Differences between managers and line employees in a quality management environment. *The international Journal of Quality & Reliability Management*, 21(2/3), 214–230.

[116] Peretti, J. (1996). *Tous DRH*. Paris: Ed. D'Organisation.

[117] Peters, T.J., & Waterman, H.R. (1987). *Na senda da excelência*. Lisboa: Publicações D. Quixote.

[118] Porter, M. (1985). *Competitive Advantage: Creating and Sustaining Superior Performance*. New York: Free Press.

[119] Porter, M. (1990). *The Competitive Advantage of Nations*. London: Macmillan.

[120] Purcell, J., & Hutchinson, S. (2007). Front-line managers as agents in the HRM-performance causal chain: Theory, analysis and evidence. *Human Resource Management Journal, 17*, 3–20.

[121] Purcell, J., & Hutchinson, S. (2007). Front-line managers as agents in the HRM-performance causal chain: theory, analysis and evidence. *Human Resource Management Journal*, 17(1), 3–20.

[122] Rego, A., & Cunha, M.P. (2009). *Liderança Positiva*. Lisboa: Edições Sílabo.

[123] Rego, A., Cunha, M.P., Costa, N.G., Gonçalves, H., & Cabral-Cardoso, C. (2006). *Gestão Ética e Socialmente Responsável: Teoria e Prática*. Lisboa: RH Editora.

[124] Ribeiro, J.L. (2003). *Estratégias de Identidade Profissional dos Gestores de Recursos Humanos: Um Estudo Exploratório*. Dissertação de Mestrado. EEG/Universidade do Minho, Braga.

[125] Russell, Z.A.; ,Steffensen D.S.; , Parker, E III., Liwen Z., Bishoff, J.D., & Ferris, G.R. (2018). High performance work practice implementation and employee impressions of line manager leadership. *Human Resource Management Review*, 28, 258–270.

[126] Schuler, R., & Jackson, S. (1997). Gestão de recursos humanos: tomando posição para o século XXI. *Comportamento Organizacional e Gestão*, 3(2), 255–274.

[127] Sculion, H., & Starkey, K. (2000). In search of the changing role of the corporate human resource function in the international firm. *International Journal of Human Resource Management*, 11(6), 1061–1081.

[128] Segalla, M., & Besseyre des Horts, C. (1998). La gestion des ressources humaines en Europe: une divergence des pratiques?. *Revue Française de Gestion*, 117, 18–29.

[129] Senge, P. (1990). *The Fifth Discipline*. New York: Currency Double-day.

[130] Shore, L.M., Chung-Herrera, B.G., Dean, M.A., Ehrhart, K.H., Jung, D.I., Randel, A.E., & Singh, G. (2009). Diversity in organizations: where are we now and where are we going? *Human Resource Management Review*, 19(2), 117–133.

[131] Siegel, D. (2009). Green management matters only if it yields more green: an economic/strategic perspective. *Academy of Management Perspectives*, 23(3), 5–16.

[132] Sikora, D.M., & Ferris, G.R. (2014). Strategic human resource practice implementation: The critical role of line management. Human Resource Management Review, 24(3), 271–281.

[133] Stainback, K., Tomaskovic-Devey, D., & Skaggs, S. (2010). Organizational approaches to inequality: inertia, relative power and environments. *Annual Review of Sociology*, 36(1), 225–247.

[134] Storey, J. (1989). *New Perspectives on Human Management*. London: Routledge.

[135] Storey, J. (1992). *Developments in the Management of Human Resources*. Oxford: Blackwell.

[136] Storey, J. (2007). *Human Resource Management: A Critical Text* (3ª Ed). London: Thomson.

[137] Strauss, A., & Corbin, J. (1998). *Basics of Qualitative Research*. London: Sage.

[138] Sully de Luque, M., Washburn, N.T., Waldman, D.A., & House, R.J. (2008). Unrequited profit: how stakeholder and economic values relate to subordinates' perceptions of leadership and firm performance. *Administrative Science Quarterly*, 53(4), 626–654.

[139] Taborda, A., & Pimentel, A. (2012). Crise. Como nos estamos a adaptar aos tempos difíceis. *In* A. Barreto (Ed.), XXI, *Dias Inquietos. Ter Opinião 2012* (pp. 8–23). Lisboa: Fundação Francisco Manuel dos Santos.

[140] Taylor, F.W. (2011). *Os Princípios da Gestão Científica* (Edição Centenária, 1911–2011). Lisboa: Edições Sílabo.

[141] Teal, T. (1996). The human side of management. *Harvard Business Review*, 74(6), 35–44.

[142] Torrington, D. (1989). Human resource management and the personnel function. *In* J. Storey (Ed), *New Perspectives on Human Resource Management* (pp. 56–66). London: Routledge.

[143] Torrington, D. (1998). Crisis and opportunity in human resources management – the challenge for the personnel function. *In* P. Sparrow & M. Marchington (Eds), *Human Resource Management – The New Agenda* (pp. 23–36). New York: Pitman Publishing.

[144] Torrington, D., & Hall, L. (1991). *Personnel Management: A New Approach*. New York: Prentice-Hall.

[145] Tregaskis, O., Heraty, N., & Morley, M. (2001). HRD in multinationals: the global/local mix. *Human Resource Management Journal*, 11(2), 35–56.

[146] Truss, K. (2001). Complexities and controversies in linking HRM with organizational outcomes. *Journal of Management Studies*, 38(8), 1121–1149.

[147] Tung, R. L. (2016). New perspectives on human resource management in a global context. *Journal of World Business*, 51, 142–152.

[148] Ulrich, D. (1987). Strategic human resource planning: why and how? *Human Resource Planning*, 10(1), 37–56.

[149] Ulrich, D. (1997a). Judge me more by my future than by my Past. *Human Resource Management*, 36(1), 5–8.

[150] Ulrich, D. (1997b). HR of the future: conclusions and observations. *Human Resource Management*, 36(1), 175–179.

[151] Ulrich, D. (1997c). *Human Resource Champions: The Next Agenda for Adding Value and Delivering Results*. Boston: Harvard Business School Press.

[152] Ulrich, D. (1998a). A new mandate for human resources. *Harvard Business Review*. 76(1): 124–134.

[153] Ulrich, D. (1998b). The future calls for change. *Workforce*, January: 77(1): 87–91.

[154] Ulrich, D., Brockbank, W., Johnson, D., Sandholtz, K., & Younger, J. (2008). *HR Competencies: Mastery at Intersection of People and Business*. Alexandria, VA: Society for Human Resource Management.

[155] Urquhart, C. (2013). Grounded Theory for Qualitative Research: A Practical Guide. London: Sage.

[156] Waeyenberg, T. V., & Decramer, A. (2018). Line managers' AMO to manage employees' performance: The route to effective and satisfying performance management. *The International Journal of Human Resource Management*, 29(22), 3093–3114.

[157] Waldman, D. A., Sully de Luque, M., & Wang, D. (2012). What can we really learn about management practices firms and countries. *Academy of Management Perspectives*, 26(1), 34–40.

[158] Welch, J. (2011). *Vencer*. Lisboa: Actual Editora.
[159] West, B. P. (2003). *Professionalism and Accounting Rules*. Abingdon: Routledge.
[160] Wheeler, A. R., Halbesleben, J. R. B., & Harris, K. J. (2012). How job-level HRM effectiveness influences employee intent to turnover and workarounds in hospitals. *Journal of Social Psychology*, 130(1), 111–113.
[161] Whittaker, S., & Marchington, M. (2003). Devolving HR responsibility to the line. Threat, opportunity or partnership. *Employee Relations*, 25(3), 245–261.
[162] Yamamoto, H. (2009). *Retention Management of Talent: A Study on Retention in Organizations*. Tokyo: Chuokeizai-sha.
[163] Yamamoto, H. (2013). The relationship between employees' perceptions of human resource management and their retention: from the viewpoint of attitudes toward job-specialties. *International Journal of Human Resource Management*, 24(4), 747–767.

# 5

# Transversal Competences: A True and Effective Support to Achieve Greater Organizational Sustainability

**André Filipe Barreira and Carolina Feliciana Machado***

School of Economics and Management, University of Minho, Portugal
E-mail: barreirandre@gmail.com; carolina@eeg.uminho.pt
*Corresponding Author

This chapter analyzes the importance of transversal competences for a human resource manager, as a support to achieve greater organizational sustainability, in small and medium-sized enterprises (SMEs).

Understanding the manager as a driver of success in SMEs and the importance that a management of competences can bring to achieve higher levels of performance and organizational sustainability, we seek in this chapter to assess the connection between these two realities, using the specificity of transversal competences, framed in what is the distinctive reality of an SME.

For this purpose, a sample consisting of students of a master's degree program in human resource management (n = 28) was used, aiming to assess the degree of evaluation of a set of 40 transversal competences based on the study developed by Silva [1].

There has been an alignment on perceptions, about the importance of certain transversal competences for a human resource manager, compared to several competences analysis studies, which supports the idea that transversal competences are in fact considered important, and may possibly constitute a valuable aid for the performance of the role of a human resource manager in an SME.

## 5.1 Introduction

This chapter has as the main aim to analyze the importance of transversal competences for a human resource manager, as a support to achieve greater organizational sustainability, in small and medium-sized enterprises (SMEs).

The question of what makes a business successful has received increasing attention from the scientific community, with particular emphasis on the manager as the main factor of success in SMEs, even in comparison with traditional factors such as differentiation of the product or difficulties in entering the market [2].

In this sense, it is important to realize that the development of competences is an issue as important as the direct provision of resources or the creation of a positive working environment [3], the reason why identification of key competences should be regarded as the initial step towards the construction of a competences management [4].

Thus, the general objective of this research is to answer the question: "What transversal competences are most important for a human resource manager in an SME?"

In order to achieve this objective, a first step is to describe what an SME is, as well as its distinctive characteristics, as regards HRM, in relation to a large company, and then analyze the importance of management competences, and in particular transversal competences and their applicability in an SME.

In the second phase, and after enunciating the methodology adopted for the research, we present the results obtained by analyzing, discussing and comparing the results with other related studies. Finally, some final considerations about the perceived understandings of the importance of transversal competences for a human resource manager are highlighted.

## 5.2 Small and Medium-sized Organizations

The Portuguese business network consists essentially of SMEs, with studies that indicate that around 97.2% of businesses in Portugal, and 55.4% of jobs are employed by micro and small enterprises [5]. SMEs have been characterized for a few decades as organizations with low-skilled, family-owned and few management practices, but this image has substantially changed with the development of several technological organizations with ability to thrive globally, investing in quality practices [5].

The image of an SME is essentially defined by the image of its manager, since it is in this one that its values, philosophies and management styles are portrayed [6]. Yet it also portrays the lack of managerial skills, in which the SME image is severely impaired, since, and as Temtime and Pansiri [7] argue, all the problems of an SME are essentially management problems, with those influenced by our own skills, dedication and effort.

In this sense, Temtime and Pansiri [7] argue that managers should be self-efficacious, which in itself requires certain competences, not only to establish organizational goals, but also to define strategies to achieve them, taking into account the specificity of the SME continuous dynamic environment, which obliges the entrepreneur to have agility and management skills.

**Distinctive characteristics**

The classical idea that SMEs should be managed in the same way as large organizations, but on a smaller scale, alone, creates a lack of understanding about SMEs that have not shown positive results [8].

As Cardon and Stevens [8] point out, all SMEs have some form of an HRM, albeit an informal one, although in general, they do not have professional human resource managers, which is only solved when the size of the SME exceeds 100 employees, which has a positive impact on the performance of the organization.

Although some authors point out that HRM practices are informal and tend to be ad hoc in SMEs [5, 9], we can still make a brief characterization about them, based on Cardon and Stevens [8], who highlight that recruitment and selection in SMEs are one of the most important and frequent practices, although, and in a general way, there does not exist a well-defined strategy using adaptation to meet the needs. As for training and development concerns, they refer to the fact that it is not structured, and is often informal, despite the notion that it is something important, a result of the constant need for change. Finally, performance appraisal is something that is not normally formalized and as such the awards and rewards are variable depending on the life cycle of the company.

Melo and Machado [5] observed the best SMEs in Portugal trying to understand the role of HRM in SMEs in a national context, reaching the conclusion that HR are still not seen as a strategic resource for organizations and as such HRM practices are not integrated with the organization's strategy.

The authors found that even informally all SMEs have some form of HRM, as well as that the existence of formal HRM practices do not depend on the size of the organization and that HRM practices are usually ad hoc and not integrated with the business strategy.

In another perspective, Barrett and Meyer [10] found that the longer the owner (who assumes management functions in the SME) assumes the responsibility of the company, the greater the importance given to the people and the way in which they should be managed, which highlights the important role that HRM should play in an SME.

However, it is important to refer to one of the problems pointed out by Melo [6] as regards SMEs, which is due to the lack of competences of the owners, since the management is carried out by people with different training areas, being, for example, usual engineers perform the functions of management. This issue will influence the whole organization, whether in the strategic part of management or in the management of the people that we look to approach in this study.

## 5.3 Competences Management

The concept "competences management" today has a wide meaning in the academic and scientific literature that has led to a great diversity of meanings and conceptions that are supported in different methodologies and with different results [11, 12].

Starting from the study of Ceitil [4], it is important, since the beginning, to understand that there are different perspectives in the approach to the concept competence that can effectively lead to different understandings. For instance, we can understand competences as attributions, qualifications, personal characteristics and behaviors.

Competences seen as attributions are used essentially in an institutional environment to designate the responsibilities and knowledge inherent to the exercise of certain functions, regardless of the person using them or not.

Competences understood as qualifications refer to the need to have certain qualifications (e.g., courses/training) to perform a certain function.

Competences as personal characteristics are those that derive essentially from the study developed by David McClelland [13], where it is argued that competences are an interconnected set of intrapersonal characteristics. That is, factors inherent in the personality of each individual, where each person can present certain traits independently of their concrete behaviors, favoring in this perspective the use of measuring instruments, based on the use of psychological tests to evaluate the individuals' psychological characteristics and personality traits, in order to identify traits or characteristics that present a high predictive validity regarding the possibility of developing future behaviors.

Finally, the perspective that evidences competences as behaviors or actions that are based on the previous perspective, but with a distinctive particularity, to the extent that is valued, is not the presence of such traits in a person, but their practical results. In other words, a competence exists only if it presents practical results, considering that "traits and characteristics are realities in potency, competences are realities in act, and as such, visible, observable, and, of course, more easily measurable" ([4], p. 34).

In what concerns the competences components usually considered, we can highlight four levels in personal terms: (a) know; (b) know-how; (c) know-be and (d) want-to-do, as well as a level of support connected with "power-making" [4].

One of the conceptual models most used to define competences is the *iceberg model* [14]. In this model, two ways of seeing competences are considered: (a) At a more internal level where the person's motivations are present: the personality traits, the self-concept and its values, less visible aspects of the competences and which are more difficult to change, corresponding to the hidden part of the iceberg; (b) And a visible top of the iceberg that translates into knowledge and skills, which represent more objectively the individuals' performance, which are seen as easier to change.

Rego et al. [12] state that among the diverse concepts of competence, and in a summary, few common points can be identified, namely:

(a) the observation about what people really do to be successful is the best way to understand the performance (even if we consider personality traits);
(b) competences can be learned and developed over time, contrary to personality traits;
(c) competences should be related to meaningful outcomes that describe desirable behaviors, not concepts that are difficult to operationalize.

In what concerns HRM, it is important to understand the concept of competences, as defended by the American School (where competences are seen as attributes), contrary to the British School (where competences are seen as behaviors) since they can be used in different HRM practices, such as recruitment and selection, training and development or even performance appraisal. However, this way of looking human resources presupposes a functionalist adoption of the organization in what concerns the definition of behaviors necessary for the performance optimization in a given function, allowing this way, to the HR manager, to objectify the decisions taken regarding, for example, recruitment and selection, definition of training plans or compensation systems [12].

Therefore, for HRM, competences management is commonly associated with a technical tool used in recruitment, training, performance appraisal and reward systems [11]. At this level, it has, as its critical distinctive feature, the fact that competences differ from the traditional description of the function analysis considering that while this one presents the tasks of a given function in details; competences constitute the ways in which these activities must be performed in order to obtain superior levels of performance [14].

It should also be mentioned that competences are changeable and can be apprehended and improved through our own experience or training [13], which gives HRM an even greater significance.

In terms of advantages for HRM, Ceitil [4] argues that the organization benefits both in terms of employee selection, strategic planning and the creation of a more participatory and motivating environment with result orientation, which lead to a greater competitiveness, commitment and differentiation. Indeed, competences management demands a very close relation between the training and productive systems, since this latter has the responsibility to identify the competences required in the individual to obtain better performances. Training then has an important role of interaction as this should be understood as a continuous process that must take place during the entire career of the employees.

In short, we can say that a competency-based human resource management system can not only enhance human resource planning and the implementation of diverse HRM plans and practices, but also the evaluation of the obtained results.

Nevertheless, we should look at the adoption of competency-based models with caution, as some implementation difficulties are also pointed out, namely in the transition from a function-based system, which has always been considered as the best way to structure an organization for a competency-based system [12].

### 5.3.1 Transversal Competences

Woodruffe [15] makes a distinction between the types of competences by distinguishing them in two broad groups: (a) technical competences and (b) transversal competences.

Technical competences refer essentially to the set of specific competences necessary to each professional function performance, and transversal competences, also called as generic or universal, with a more comprehensive focus, encompassing the totality of professional functions.

As highlighted by Ceitil [4], transversal competences are characterized by not being specific to a given context as well as by some common characteristics such as: (a) multi-functionality; (b) transferability; (c) based on cognition; (d) multidimensionality; (e) learning and (f) comprehensiveness.

The number of different categorizations of transversal competences is measured according to the number of authors who proposed to study them [16]. However, it seems important to highlight a European-wide project that sought to harmonize educational structures in Europe where several competences, generic and specific, associated with different areas, were analyzed, Tuning Project, where three types of competences are considered [17]:

(a) instrumental competences, closely linked with cognitive, technological, methodological and linguistic abilities;
(b) interpersonal competences, which deal with social competences of relationship;
(c) systemic competences that interconnect knowledge, understanding and sensitivity in order to perceive how realities relate to and influence each other.

Silva [1], in a study focused on the Portuguese reality, identified, based on several authors, a set of 40 transversal competences in which he sought to assess the degree of importance both for graduates and for employers of various professional areas. It was also based on this study that we defined the competences that we would analyze and which are explained in Annex A.

To HRM professionals, Ulrich et al. [18], in their research, sought to define which competences are the most valuable for all stakeholders in an organization, reaching to a model based on six domains that focus on organizational relationships, processes and capacities.

In what concerns processes, they point out that an HR manager should have a more operational perspective with regard to people's management policies, including administrative needs, which should be ensured as a way of ensuring the consistent credibility of policies, as well as a more business-related perspective where it is necessary to establish goals and objectives to face the opportunities and threats of the external environment and social context.

As for organizational capacities, an HR manager should be:

(a) a change agent, considering that he must respect the culture of the organization but seek to shape a new culture; at the same time, he must be a dynamic person and a change facilitator;
(b) a talent manager who ensures the alignment of strategic needs with organizational capabilities, focusing on how individuals enter and grow within the organization;
(c) strategy architect, as he should have a vision of how the organization can gain in the future by playing an active role in fulfilling this vision, recognizing business trends and predicting possible obstacles to success, linking internal organization to customer expectations outside.

Finally, Ulrich et al. [18] point out that a human resource manager must, above all, be someone who is credible (respected, admired and heard) and active (having a point of view, taking a position and defying assumptions). He must possess, necessarily, these two characteristics in order to be able to sustain all of the other competences listed above, which highlight the dimension of the transversal competences that a human resource manager must possess, regardless of the type of organization where it is inserted.

Ulrich et al. [18] consider that each organization should identify high- and low-performing employees and interview them about the critical aspects of their performance. It must be done to determine which values, knowledge or skills differentiate them in order to develop the competences that lead to a high performance. With the right competences, HR managers are more likely to be able to engage employees. This is a crucial issue for HRM.

In this sense, Cassidy [19] also refers to the importance of transversal competences for individuals in the business reality, relating them to the performance of managers and drawing attention to the need to determine if they are effectively useful for professional performance, since the education system often ignores transversal competences.

### 5.3.2  Competences Management in a SME

The CEO/manager plays a central role for the SME performance, being essential to the business development and its increasing performance [2].

Man, Lau and Snape [20] also consider that the influence of a manager in an SME is critical, with competency management playing an essential role. They demonstrate that although the perceived differences about the environment and company resources play an important role in the creation of organizational competitiveness, it cannot be developed without some entrepreneurial competences, such as relationship, innovative, opportunity and human competences, contributing, as well, the entrepreneurial competences to the long-term success of the company performance.

As we have seen, competences may be related with the achievement of superior performance. In HRM field, it can be closely related with recruitment and selection processes, training and development, career management or compensation systems [12]. However, it requires an HRM integrated action in an organization, insofar as, as it promotes a common understanding based on what competences related to the functions and structural levels are, both of authority and responsibility, it also promotes the performance appraisal management based on observable behaviors. Increasing, this way, behavioral predictability based on past behaviors allows a comparison between the function profile and the behaviors' profile of the function holder.

Although there are no studies that specifically relate the importance of transversal competences to a human resource manager in an SME, some authors have already tried to study the impact of competences for a manager in this type of organization.

Man et al. [3] carried out a survey about the entrepreneurial competences with relevance in an SME context, concluding about the importance of opportunity, relationship, conceptual, organizational, strategic and commitment competences.

Wagener, Gorgievski and Rijsdijk [21], looking to identify critical competences to the entrepreneur' success, regardless of the sector or business branch, found that "taking risks" seems to play an important role both in life and in the success of a manager. However, this issue is also closely linked with failures and setbacks, being crucial that entrepreneurs have the necessary competences to deal with these risks and possible negative consequences.

Kyndt and Baert [22] also sought to survey competences, essential to the entrepreneur success in small and medium-sized enterprises, finding the following:

a) perseverance; b) self-knowledge; c) learning orientation; d) knowledge of potential returns on investment; e) determination; f) planning for the future; g) independence; h) network construction; i) ability to persuade; j) seeking opportunities; k) market awareness; l) environmental and social awareness. From these competences, Kyndt and Baert [22] analyzed all of them in 34,968 aspiring entrepreneurs. After 3–5 years, they evaluated 3239 of them, concluding that the competences that made them truly entrepreneurs were perseverance and the perception of the market needs.

It is important to note that McClelland [13] also warned against this by pointing out that it is essential to be perseverant in order to continuously face the obstacles and difficulties that appear throughout the career.

Alves [23], when assessing the perception of the critical transversal competences in an SME in the Portuguese context, found that the most mentioned by the interviewees were: a) responsibility, b) team spirit, c) autonomy, d) creativity and finally (e) polyvalence.

Overall, to the creation of a competency-based system, Ceitil [4] states that it will be necessary to follow five phases, which can be transposed to an SME, namely: a) identification of key competences; b) portfolio description of key competencies; c) competences assessment; d) definition of competences development action plans; e) competences development assessment.

In this study, we decided to focus on the first two phases, identifying up to this point a set of transversal competences considered essential. In the next chapters, we will try to objectify this analysis on competences to obtain superior performance in an SME through the perception of students of a master's degree in human resource management.

## 5.4 Methodology

Research methodology used in this work was based mainly on the quantitative method where a questionnaire survey was developed to evaluate the perception of the importance of a set of 40 (transversal) competences based on Silva's research [1].

The target population to this research were the first-year students of master's degree in human resource management at a Portuguese university[1].

The questionnaire survey was disseminated online[2] between April 20 and 30, 2017, through the page of the Facebook group that the master's students in human resource management possess, aiming to reach the largest number of students given the constant use of the group by the vast majority of students. In this approach, it was explained the purpose of the research as well as what was intended with the collaboration. In particular, they were asked to classify according to their perception and by the degree of importance, the

---

[1]About 47 students considering the number of enrolled in most of the curricular units and in the Facebook group (the withdrawals are not counted in the first semester or students who are repeating units and were already enrolled in other years).

[2]Through the link https://www.survio.com/survey/d/C2Y7X1D5C6M1D9W6G.

competences that a human resource manager must possess in a SME to obtain superior performances.

The questionnaire consisted of four initial questions aiming to characterize the sample, namely gender, age, whether they have already performed functions in the area of human resource management and how long have they performed these functions.

A 5-point Likert scale was also used to gauge the importance given to each transversal competency, being numbered as follows: "1 – Not important", "2 – Few important", "3 – Important", "4 – Very important "and" 5 – Extremely important".

## 5.5 Results Analysis and Discussion

In this section, we present the results obtained from a descriptive analysis of the variables previously listed in the questionnaire survey.

The used sample is composed of 24 female and 4 male students.

About the participants' age, it is mostly from 23 to 27 years, with 5 students over 32 years of age, 3 with less than 22 years of age and only 2 with age between 28 and 32 years (Table 5.1).

**Table 5.1** Socio-demographic data: Gender and age

| Gender | | | |
| --- | --- | --- | --- |
| Female – 86% | | Male – 14% | |
| Age | | | |
| Less than 22 years – 11% | 23–27 years – 64% | 28–32 years – 7% | Up to 32 years – 18% |

From the 28 participants in the study, 18 never developed functions in the HRM area and 10 have developed at some point. Of these, 8 worked in a company with more than 250 employees and only 2 in an average company (Table 5.2).

**Table 5.2** Already worked in HRM area (Issue 3)

| | |
| --- | --- |
| No | 67% |
| Yes, in a micro organization (<10 employees) | 0 |
| Yes, in a small organization (<50 employees) | 0 |
| Yes, in a medium organization (<250 employees) | 6% |
| Yes, in an organization with more than 250 employees | 27% |

In question 4 "How long did you perform HRM functions?", we have observed that 50% of the respondents performed functions less than 1 year while the other 50% developed HRM functions from 2 to 5 years. No one had performed these functions more than 6 years.

The last section of the survey aimed to understand which transversal competences HRM students consider an HR manager should possess, to perform their functions in an SME in order to obtain higher performances. This is listed in Table 5.3 for the entire sample.

From Table 5.3, we can see that the competence perceived as most important is problem solving, followed by competences related to oral communication, adaptation to change, listening ability and critical thinking.

At the other extreme, competences less valued are numeracy, the only one whose average was below 3 points on the Likert scale (1 to 5), followed by personal presentation, general culture, foreign languages and written communication.

In order to obtain a comparison with the perception between students with and without professional experience, we considered as critical to verify what was the perception expressed by the students' sample with professional experience in the HRM field. We observed a similarity in what concerns the TOP10 of the competences understood as critical to the achievement of superior performances, and for these, only the persistence and the competence to live with multiculturalism differ from the most important ones. With regard to less valued competences, we also have numeracy, general culture and personal presentation as the least valued.

Compared with the competences listed in the World Economic Forum (2016), and considered as the competences that will be most valued in the labor market in 2020, we find some similarities in relation to the valuation of the type of competences, namely in solving problems that are positioned in the first place in both analyzes, which reflects an issue really important to be taken into account, as well as other equally valued competences such as critical thinking, planning and organization, as well as others that may be intertwined with emotional intelligence such as the ability to listen, interpersonal relationship and self-control that were evaluated in this survey instead of emotional intelligence but that in its genesis has these components.

Silva [1], with another study at the Portuguese national level also obtained some results that can be compared in terms of importance, both in the assessment that graduates assigned to certain competences, and the employers, specifically in problem solving competences, interpersonal relationship, planning/organization and adaptation to change that are coincident with the results found in the present analysis.

From this research, the most highlighted competences in terms of valuation difference are negotiation and the development of others, which in the study of Silva [1] are the 39th and 36th competences less valued, while in

**Table 5.3** Descriptive analysis of transversal competences

| Competences | Variable Code | N | Minimum | Maximum | Mean | Standard Deviation |
|---|---|---|---|---|---|---|
| Problem solving | c6 | 28 | 3.00 | 5.00 | 4.8571 | 0.52453 |
| Oral communication | c2 | 28 | 3.00 | 5.00 | 4.7857 | 0.56811 |
| Change adaptation | c10 | 28 | 4.00 | 5.00 | 4.7857 | 0.41786 |
| Ability to hear | c26 | 28 | 4.00 | 5.00 | 4.7857 | 0.41786 |
| Critical sense | c16 | 28 | 3.00 | 5.00 | 4.7143 | 0.59982 |
| Interpersonal relationship | c27 | 28 | 3.00 | 5.00 | 4.6429 | 0.73102 |
| Ethical compromise | c17 | 28 | 3.00 | 5.00 | 4.6071 | 0.73733 |
| Conflict management | c36 | 28 | 3.00 | 5.00 | 4.6071 | 0.62889 |
| Self-control | c33 | 28 | 4.00 | 5.00 | 4.5357 | 0.50787 |
| Planning/organization | c14 | 28 | 3.00 | 5.00 | 4.5357 | 0.63725 |
| Decision making | c34 | 28 | 3.00 | 5.00 | 4.5000 | 0.63828 |
| Innovation/creativity | c11 | 28 | 3.00 | 5.00 | 4.4643 | 0.63725 |
| Live with multiculturalism/ diversity | c15 | 28 | 3.00 | 5.00 | 4.4643 | 0.69293 |
| Persistency | c32 | 28 | 4.00 | 5.00 | 4.4286 | 0.50395 |
| Motivation | c35 | 28 | 3.00 | 5.00 | 4.4286 | 0.79015 |
| Self-confidence | c20 | 28 | 3.00 | 5.00 | 4.3929 | 0.73733 |
| Negotiation | c29 | 28 | 3.00 | 5.00 | 4.3571 | 0.73102 |
| Other's development | c40 | 28 | 3.00 | 5.00 | 4.3571 | 0.78004 |
| Stress tolerance | c19 | 28 | 3.00 | 5.00 | 4.3214 | 0.72283 |
| Other's motivation | c37 | 28 | 3.00 | 5.00 | 4.3214 | 0.66964 |
| Leadership | c12 | 28 | 3.00 | 5.00 | 4.3214 | 0.81892 |
| Availability to continuous apprenticeship | c22 | 28 | 3.00 | 5.00 | 4.2857 | 0.59982 |
| Planning – action | c28 | 28 | 3.00 | 5.00 | 4.2857 | 0.71270 |
| Initiative | c31 | 28 | 3.00 | 5.00 | 4.2500 | 0.58531 |
| Networks establishment | c38 | 28 | 3.00 | 5.00 | 4.2500 | 0.75154 |
| Ability to ask | c25 | 28 | 3.00 | 5.00 | 4.2500 | 0.75154 |
| Autonomy | c9 | 28 | 3.00 | 5.00 | 4.2143 | 0.83254 |
| Influence/persuasion | c24 | 28 | 3.00 | 5.00 | 4.2143 | 0.83254 |
| Risk assumption | c39 | 28 | 3.00 | 5.00 | 4.2143 | 0.56811 |
| Team work | c4 | 28 | 3.00 | 5.00 | 4.2143 | 0.78680 |
| Customer orientation | c5 | 28 | 3.00 | 5.00 | 4.0000 | 0.86066 |
| Attention to detail | c23 | 28 | 3.00 | 5.00 | 3.9643 | 0.69293 |
| Sensitivity to business | c18 | 28 | 2.00 | 5.00 | 3.7857 | 0.95674 |
| ICT | c1 | 28 | 2.00 | 5.00 | 3.6786 | 0.72283 |
| Collection and processing of information | c13 | 28 | 3.00 | 5.00 | 3.6786 | 0.72283 |
| Written communication | c3 | 28 | 2.00 | 5.00 | 3.6429 | 1.28277 |
| Foreign languages | c8 | 28 | 2.00 | 5.00 | 3.6071 | 0.83117 |
| General culture | c21 | 28 | 2.00 | 5.00 | 3.4643 | 0.88117 |
| Personal presentation | c30 | 28 | 2.00 | 5.00 | 3.1071 | 0.87514 |
| Numeracy | c7 | 28 | 2.00 | 4.00 | 2.8929 | 0.73733 |
| Final mean | Valid N | 28 | 2.93 | 4.98 | 4.23 | 0.71 |

our analysis, they are in 17th and 18th. As well as in the opposite direction are the information and communication technologies, in 33rd in our analysis, being the 3rd most valued in the research of Silva (2008). These differences in terms of valuation do not seem to be out of phase with what is reality, since the valuation result of the study scope of Silva [1], had only a sample of 7 respondents in a total of 421 of the management area (1.7%), in comparison with our entire sample of human resource management students, where greater emphasis is placed on negotiation and development of others competences and not so much on competences related to information and communication technologies.

We can also compare the results of the present research with those obtained in the study of competences of Ulrich et al. [18] that focused on human resource managers in particular. Ulrich et al. [18] verified that a human resource manager, in addition to the technical competences that he/she should have and that were analyzed in Section 5.3, should be a "credible activist" from which should be included competences related to the credibility of a human resource manager, where he must be respected, admired and hear, but also must be active, have a critical point of view, assume a position and challenge assumptions, competences that, together, predict the effectiveness of an HRM.

This set of competences (Table 5.4) can be framed in oral communication, critical thinking, interpersonal relationships, ethical commitment, decision making and persistence, competences that in our study are the most valued, demonstrating a convergence in the degree of importance that is given to competences considered more essential, in this case specifically to a human resource manager.

**Table 5.4**   Descriptive analysis of transversal competences of students with experience in HRM

| | | N | Minimum | Maximum | Mean | Standard Deviation |
|---|---|---|---|---|---|---|
| Problem solving | c6 | 10 | 5.00 | 5.00 | 5.0000 | 0.00000 |
| Change adaptation | c10 | 10 | 5.00 | 5.00 | 5.0000 | 0.00000 |
| Ethical compromise | c17 | 10 | 4.00 | 5.00 | 4.9000 | 0.31623 |
| Critical sense | c16 | 10 | 4.00 | 5.00 | 4.8000 | 0.42164 |
| Self-control | c33 | 10 | 4.00 | 5.00 | 4.7000 | 0.48305 |
| Oral communication | c2 | 10 | 3.00 | 5.00 | 4.6000 | 0.84327 |
| Planning/organization | c14 | 10 | 4.00 | 5.00 | 4.6000 | 0.51640 |
| Ability to hear | c26 | 10 | 4.00 | 5.00 | 4.6000 | 0.51640 |
| Persistency | c32 | 10 | 4.00 | 5.00 | 4.5000 | 0.52705 |
| Live with multiculturalism/ diversity | c15 | 10 | 3.00 | 5.00 | 4.4000 | 0.69921 |

## 5.6  Final Remarks

Considering that an organization in order to become sustainable and have a long and responsible participation in its markets needs a competent labor force, in this study, we have explored the importance of competency management, taking into account the specificities of SMEs.

We can see that competences can be defined as an integrated set of knowledge, skills and attitudes that are changeable and possible to be apprehended and improved and can be used to achieve higher levels of performance [13, 14].

To HRM, competency management can not only enhance human resource planning, and the implementation of various HRM plans and practices, but also the evaluation of the obtained results [4, 12].

As such, the importance of this research to an SME lies essentially in raising the perspective of a future human resource manager about the competences understood as essential to ensure superior performance in the role of the human resource manager, which will always be a starting and comparing point with what is desired by the organization.

In what concerns the key question established at the beginning of this study, namely, "What transversal competences are most important for a Human Resource Manager in an SME?", we have observed that there is indeed a positive alignment about the competences necessary in the labor market, supported either by Ulrich et al. [18] and Silva [1], and recently by the World Economic Forum [24], in particular with regard to problem solving, critical thinking, planning and organization as well as some transversal competences related to interpersonal relationships such as oral communication, listening skills and self-control, which end up being very much in line with what is expected of a human resource manager – to perceive others and make themselves understood.

We have also concluded that at HRM level, transversal competences are, in general, perceived as very important, with an alignment in the perception of future human resource managers with those that are considered by some studies as essential competences for a performance of excellence.

However, it should be borne in mind that competences models and the systems derived from them are not a panacea for all organizational problems and challenges, particularly by the associated costs, to profound changes in organizational architecture [12]. Even so, the development of a competence matrix can be a very important issue in an organization, since it is adjusted to the organization's strategy and could be based on specific competences according to concrete needs and not in standardized models.

For SMEs in particular, we must take into account the previously mentioned factors, as well as the specificities of these organizations, which may negatively influence the implementation of this type of models, in particular because they do not have professional human resource managers [8], and have ad hoc HRM practices that are not integrated with the business strategy [5].

However, we must also assess SME's needs. In particular, the owners' lack of competences, as well as the fact that management is carried out by people with diverse training areas [6], issues that will influence the whole company including the management of people. And if we understand, as Man et al. [20], that the influence of a manager in an SME is indeed critical, the issue of defining the competences needed to achieve better performance is undoubtedly crucial to the SME's success, as it allows not only managers to identify transversal competences able to achieve organizational performance requiring a development, as well as to the whole structure of university and business training on the way forward in the development of future human resource managers.

In short, we believe that it is possible to improve systems and tools for the management and development of human resources in organizations, regardless of their type, even though we are aware that we will never be able to achieve total objectivity in these processes.

In order to continue this research in future studies, it would be important to assess the real influence that these competences have on the effective performance of an SME HR manager, as well as its prevalence on technical competences. This would be essential in assessing the competences to be sought in recruitment and selection, as well as those that should be developed in a training and development process and in performance appraisal.

## References

[1] Silva, P. (2008). Competências transversais dos licenciados e sua integração no mercado de trabalho (Dissertação de Mestrado). Universidade do Minho, Braga.

[2] Wiklund, J., & Shepherd D. (2005). Entrepreneurial Orientation and Small Business Performance: a Configurational Approach, *Journal of Business Venturing 20*(1), 71–91. https://doi.org/10.1016/j.jbusvent.2004. 01.001

[3] Man, Y., Lau, T., & Chan. F. (2002). The Competitiveness of Small and Medium Enterprises: A Conceptualization with Focus on Entrepreneurial Competencies. *Journal of Business Venturing, 17*(2), 123–42. https://doi.org/10.1016/S0883-9026(00)00058-6

[4] Ceitil, M. (2007). *Gestão e desenvolvimento de competências.* Lisboa: Edições Sílabo.

[5] Melo, P. & Machado, C. (2013). Human resource management in small and medium enterprises in Portugal: rhetoric or reality?, *International Journal of Entrepreneurship and Small Business, 20*(1), 117–134. https://doi.org/10.1504/IJESB.2013.055696

[6] Melo, P. (2015). A Gestão de Recursos Humanos nas Pequenas e Médias Empresas em Portugal: Definição de um Modelo de Atuação (Tese de Doutoramento). Universidade do Minho, Braga.

[7] Temtime, Z. T., & Pansiri, J. (2006). Perceived managerial problems in SMEs: evidence from Botswana, *Development and Learning in Organizations, 20*(5), 15–17. https://doi.org/10.1108/14777280610687998

[8] Cardon, M., & Stevens, C. (2004). Managing human resources in small organizations: What do we know?. *Human Resource Management Review, 14*, 295–323. https://doi.org/10.1016/j.hrmr.2004.06.001

[9] Wagar, T. (1998). Determinants of human resource management in small firms: some evidence from Atlantic Canada, *Journal of Small Business Management, 36*(2), 13–23.

[10] Barrett, R., & Meyer, M. (2010). Correlates of perceiving human resource management as a 'problem' in smaller firms, *Asia Pacific Journal of Human Resources*, *48*(2), 133–150. https://doi.org/10.1177/10384 11110368464

[11] Burgoyne, J. (1993). The Competence Movement: Issues, Stakeholders and Prospects. *Personnel Review, 22*(6), 6–13. https://doi.org/10.1108/EUM0000000000812

[12] Rego, A., Cunha, M. P., Gomes, J., Cunha, R. C., Cabral-Cardoso, C., & Marques, C. A. (2015). *Manual de gestão de pessoas e do capital humano.* (3a ed). Lisboa: Edições Sílabo.

[13] Mclelland, D. (1973). Testing for Competence Rather Than for Intelligence. *American Psychologist, 28*, 1–4. https://doi.org/10.1037/h0034092

[14] Spencer, L. M., & Spencer, S. M. (1993). *Competence at work: models for superior performance.* New York: John Wiley & Sons.

[15] Woodruffe, C. (1993). What is Meant by a Competency. *Leadership & Organization Development Journal*, *14*(1) 29–36. https://doi.org/ 10.1108/eb053651

[16] Santos, M. (2016). Competências transversais na formação dos alunos de Economia e Gestão (Dissertação de Mestrado). Universidade do Minho, Braga.

[17] González, J., & Wagenaar, R. (2003). *Tuning: Educational Structures un Europe. Final Report – Phase One.* University of Deusto and University of Groningen. Disponível em http://tuningacademy.org/wp-content/uploads/2014/02/TuningEUI_Final-Report_EN.pdf, accessed in May 2017.

[18] Ulrich, D., Brockbank, W., Johnson, D., & Younger, J. (2007). Human Resource Competencies: Responding to Increased Expectations, *Employment Relations Today*, *34*(3), 1–12. https://doi.org//10.1002/ert. 20159

[19] Cassidy, S. (2006). Developing Employability Skills: Peer Assessment in Higher Education. *Education and Training*, *48*(7), 508–517. https://doi.org/10.1108/00400910610705890

[20] Man, Y., Lau, T., & Snape, E. (2008). Entrepreneurial Competencies and the Performance of Small and Medium Enterprises: An Investigation through a Framework of Competitiveness. *Journal of Small Business & Entrepreneurship, 21*(3), 257–276. https://doi.org/10.1080/08276331.2008.10593424

[21] Wagener, S., Gorgievski, M., & Rijsdijk, S. (2010). Businessman or host? Individual differences between entrepreneurs and small business owners in the hospitality industry. *The Service Industries Journal*, 30, 1513–1527. https://doi.org/10.1080/02642060802624324

[22] Kyndt, E., & Baert, H. (2015). Entrepreneurial competences: Assessment and predictive value for entrepreneurship. Journal of Vocational Behavior, *90*, 13–25. https://doi.org/10.1016/j.jvb.2015.07.002

[23] Alves, M. (2016). Conceção de um modelo de avaliação de desempenho por competências numa PME (Dissertação de Mestrado). Universidade do Minho, Braga.

[24] World Economic Forum. (2016). *The Future of Jobs.* Geneva; World Economic Forum. Disponível em http://www3.weforum.org/docs/WEF_ Future_of_Jobs.pdf

# Index

# About the Editor

**Carolina Machado** received her PhD in management sciences (Organizational and Policies Management Area/Human Resources Management) from the University of Minho in 1999, master's degree in management (Strategic Human Resource Management) from the Technical University of Lisbon in 1994, and degree in business administration from University of Minho in 1989. Teaching in the human resource management subjects since 1989 at the University of Minho, she is Associate Professor since 2004, with experience and research interest areas in the field of human resource management, international human resource management, human resource management in SMEs, training and development, emotional intelligence, management change, knowledge management and management/HRM in the digital age. She is Head of the Department of Management and Head of the Human Resources Management Work Group at University of Minho, as well as Chief Editor of the International Journal of Applied Management Sciences and Engineering (IJAMSE), guest editor of journals, books editor and book series editor, as well as reviewer in different international prestigious journals. In addition, she has also published both as editor/co-editor and as author/co-author several books, book chapters and articles in journals and conferences.